Reviews of

109 Physiology, Biochemistry and Pharmacology

Editors

P. F. Baker, London · H. Grunicke, Innsbruck
E. Habermann, Gießen · R. J. Linden, Leeds
P. A. Miescher, Genève · H. Neurath, Seattle
S. Numa, Kyoto · D. Pette, Konstanz
B. Sakmann, Göttingen · W. Singer, Frankfurt/M
U. Trendelenburg, Würzburg · K. J. Ullrich, Frankfurt/M

With 14 Figures

Springer-Verlag Berlin Heidelberg GmbH

ISBN 978-3-662-31063-2   ISBN 978-3-540-47780-8 (eBook)
DOI 10.1007/978-3-540-47780-8
Library of Congress-Catalog-Card Number 74-3674

© by Springer-Verlag Berlin Heidelberg 1987
Originally published by Springer-Verlag Berlin Heidelberg New York in 1987.
Softcover reprint of the hardcover 1st edition 1987

Typesetting: K + V Fotosatz, Beerfelden

2127/3130-543210

Contents

Indexed in Current Contents

Rev. Physiol. Biochem. Pharmacol., Vol. 109
© by Springer-Verlag 1987

Mucosal Innervation and Control of Water and Ion Transport in the Intestine

JANET R. KEAST

Contents

Department of Anatomy and Histology, Flinders University of South Australia, Bedford Park
S.A. 5042, Australia

Abbreviations: A, adrenaline; ACh, acetylcholine; ATP, adenosine 5'-triphosphate; cAMP, adenosine 3',5'-cyclic monophosphate; CCK, cholecystokinin; CGRP, calcitonin gene-related peptide; ChAT, choline acetyltransferase; CT, cholera toxin; DMPP, 1,1-dimethyl-4-phenylpiperazinium; DYN, dynorphin; EFS, electrical field stimulation; ENK, enkephalin; ESP, excitatory synaptic potential; G, transmembrane conductance; GRP, gastrin-releasing peptide; 5-HT, 5-hydroxytryptamine (serotonin); IR, immunoreactive/immunoreactivity; I_{sc}, short-circuit current; ISP, inhibitory synaptic potential; NA, noradrenaline; NPY, neuropeptide Y; 6-OHDA, 6-hydroxydopamine; PD, potential difference; PG, prostaglandin; PHI, peptide histidine isoleucine; PP, pancreatic polypeptide; PYY, peptide YY; QNB, quinuclidinyl benzilate; SOM, somatostatin; SP, substance P; ST, *Escherichia coli* heat-stable enterotoxin; TTX, tetrodotoxin; VIP, vasoactive intestinal peptide

---

1 Introduction

Extensive networks of nerve fibres are found in the muscle and mucosa in the intestine. Most of these fibres arise from ganglion cells in the myenteric or submucous plexuses, whereas a relatively small proportion arise from extrinsic ganglia. Although considerable attention has been paid to the innervation of the external musculature and the control of motility, until the past few years comparatively little interest has been shown in the distribution and roles of nerve fibres in the mucosa. A variety of functions may be subserved by these nerve fibres. Sensory nerve fibres in the mucosa are responsible for detecting chemical, osmotic and mechanical stimuli, while motor functions potentially regulated by mucosal nerves include contraction of the muscularis mucosae, secretions from endocrine and goblet cells, vascular tone, and movement of water, ions and nutrients across the epithelium. It is also possible that some nerve fibres have a trophic function.

Of these many possible functions, only the control of epithelial water and ion movement will be discussed in this review. Although some evidence implicating nerves in this role dates back as far as the last century, a convincing body of data has been generated only in the past few years. Some of this information has recently been summarized by Tapper (1983), Hubel (1985) and Cooke (1986).

2 Distribution of Mucosal Nerve Fibres

In 1858 Billroth described a plexus of fine nerve fibres beneath the glands of the human small intestine, and since then occasional references to mucosal innervation have appeared in the literature; a number of more comprehensive descriptions of this nerve network have also been published (Breiter and Frey 1862; Goniaew 1875; Drasch 1881; Müller 1892; Berkley 1893; Ramón y Cajal 1894, 1911; Müller 1911; Sabussow 1913; Hill 1927; Oshima 1929; Waddell 1929a, b; Schabadasch 1930; Stöhr 1934, 1952; Palay and Karlin 1959; Schofield 1960; Honjin et al. 1965; Honjin and Takahashi 1966; Pick 1967; Stach 1973; Stach and Hung 1979). Each of these studies has demonstrated that the mucosa of the small intestine is provided with a rich supply of nerve fibres. Most of the early work relied on methylene blue, gold chloride or silver impregnation (Golgi) methods for staining neural tissue, and the distribution and density of mucosal innervation described in these studies have largely been consistent with recent immunohistochemical and ultrastructural work. Acetylcholinesterase localization, originally thought to label cholinergic neurons specifically, has also been used to stain a large proportion of the mucosal nerve fibres. The general impression is gained that there are no large differences between duodenum, jejunum or ileum in the degree or pattern of mucosal innervation and the description here will apply to all areas of the mammalian small intestine.

The mucosal nerve fibres form an extensive interlacing network throughout the depth of the lamina propria. This network has commonly been subdivided into the intravillus (or villus), periglandular and subglandular plexuses (Fig. 1), although these plexuses are not clearly delineated and appear to be continuous with each other. It is generally agreed that the densities of innervation around the crypts and in the villi are similar, although there have been occasional reports that either the crypts (Oshima 1929) or the villi (Stach 1973; Stach and Hung 1979) have the higher density of nerves. Such discrepancies could be due to differences in staining procedures, fixation and sectioning techniques, or species studied. The mucosal nerve fibres are unmyelinated and varicose and under the light microscope the majority of the nerve strands appear as either single fibres or small bundles; however, ultrastructural studies suggest that nerve strands which appear by light microscopy to be only single fibres are often, in fact, small bundles of fibres.

The intestinal mucosa is comprised of a complex arrangement of cell and tissue types. The loose connective tissue of the lamina propria has an extensive supply of lymphatics and blood vessels, including small arterioles, venules and capillaries; small numbers of smooth muscle fibres may also be found in the villi of some species. A simple columnar epithelium covers the mucosa, and is itself a mixture of cell types, including simple absorptive/secretory cells of the villi and crypts, goblet cells, endocrine cells, Paneth

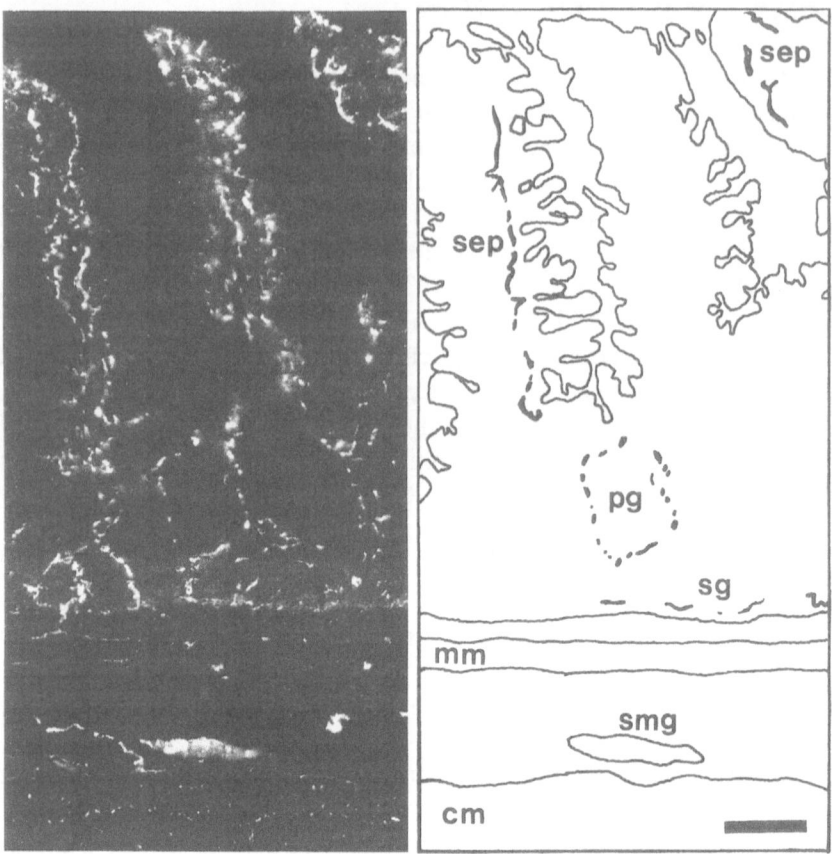

Fig. 1. Locations of nerve fibres in the intestinal mucosa. On the *left* is a micrograph of a section of dog ileum, in which mucosal nerve fibres are revealed by their immunoreactivity for VIP. The micrograph shows nerve fibres in the full thickness of mucosa (luminal surface at the *top* of the picture), submucosa (including a submucous ganglion with an immunoreactive neuron and nerve terminals) and circular muscle. On the *right* is a drawing of the main tissue layers in this micrograph, particularly to show examples of mucosal nerve fibres classified as subglandular (*sg*), periglandular (*pg*) and subepithelial (*sep*). Most of the periglandular fibres seen here could also be classified as subepithelial. The muscularis mucosae (*mm*), a submucous ganglion (*smg*) and circular muscle are indicated. Calibration bar represents 100 μm. (Micrograph kindly provided by JB Furness and M Costa)

cells and undifferentiated (stem) cells. Hence there are many possible target tissues and motor functions for the mucosal nerve fibres; sensory nerve fibres are also present in the mucosa. From the density of nerve fibres observed in the intestinal mucosa, it is not surprising that many of the cell and tissue types are closely associated with nerve fibres. Some nerve fibres travel close to the villous and glandular epithelium, while others accompany smooth muscle fibres and blood vessels. Those which are found close to blood vessels frequently encircle them.

There have been some suggestions of distinct, specific nerve plexuses for each type of target tissue in the mucosa. Berkley (1893) believed that the nerve plexuses around the crypts were separate from those in the villi, although this has not commonly been reported. The subepithelial plexus is, in some preparations, distinct from fibres supplying the mucosal muscle or vascular tissue (Drasch 1881; Müller 1892; Oshima 1929); some of the mucosal vascular nerves are clearly continuations of the nerve fibres which follow submucosal blood vessels (Schofield 1960). Stach and Hung (1979) described two distinct types of villus innervation − the majority of nerve fibres were thought to be continuous with the network in the crypts, whereas other nerve fibres appeared to travel directly from the submucosa to the villi. Although most authors agree that the subepithelial plexus is quite uniform throughout the mucosa, there has been one report of specific or preferential innervation of enterochromaffin cells (Lundberg et al. 1978). This study showed that nerve bundles located close to enterochromaffin cells (but not those close to other epithelial cells) have little or no covering by Schwann cells and are thus presumably more likely to be involved in neurotransmission at those sites. Wade and Westfall (1985) have recently reported that, in the duodenum, small bundles of unmyelinated nerve fibres partially encircled by Schwann cells were commonly present $0.5-1.0\ \mu m$ away from enterochromaffin cells.

Some of the very early studies described nerve fibres which penetrated between mucosal epithelial cells (Drasch 1881; Kuntz 1913; Müller 1921; Hill 1927). These nerve endings were found in greater numbers in the villi than around the crypts and were postulated to have a sensory function (Hill 1927). Although an ultrastructural study by Dermietzel (1971) also found a single intra-epithelial nerve ending in the pyloric mucosa, intra-epithelial nerve fibres have not been found in the intestine using recent histochemical and other microscopic techniques and, if present, are probably very rare. A number of factors could have led to the observation of structures which resembled and were falsely described as intra-epithelial nerve fibres; these include partial staining of enterochromaffin cells using silver impregnation techniques, the study of thick sections or whole mounts, and deposition of heavy metals between epithelial cells.

Ultrastructural studies have shown that most of the mucosal nerve fibres are varicose, with individual varicosities containing many vesicles; these vesicles are thought to represent transmitter stores. Close contacts between vesiculated nerve profiles and epithelial cells of the crypts and villi, smooth muscle, blood vessels and lacteals, and connective tissue elements (e.g. mast cells, macrophages) of the mucosa are common (Palay and Karlin 1959; Honjin et al. 1965; Honjin and Takahashi 1966; Pick 1967; Stach 1979; Newson et al. 1983). Although the distances between nerve fibres and other tissues may be quite small ($50-100\ nm$), junctional specializations seem to be rare (Newson et al. 1979b; Wade and Westfall 1985). However, in the autonomic nervous system, transmission occurs at multiple sites along nerve terminals, and these sites are characterized by the presence of transmitter stores (vesicles). An indication of functional innervation in the mucosa may therefore be better given by the proximity of nearby tissues and the degree of Schwann cell covering for any given nerve fibre profile, with less covering perhaps indicating more effective neurotransmission at that site. In the pylorus it has been estimated that approximately 61% of the surface of mucosal nerve fibres is not covered by Schwann cells and is presumably able to be involved in neurotransmission (Dermietzel 1971). Such calculations have not yet been made for the intestinal mucosal innervation, although the nerve fibre bundles have a similar appearance in this area.

3 Origins of Mucosal Nerve Fibres

Early light microscope studies distinguished two morphological types of mucosal nerve fibres by differences in diameter. The thicker of these is scarcer in the mucosa of the small than the large intestine and appears to arise from the vagus (Waddell 1929a, b; Schofield 1960). Several studies have indicated that some of the thinner fibres originate from sympathetic ganglia (Kuntz 1922; Hill 1927; Honjin 1951; Makino 1955; Fehér and Vajda 1974), whereas most of the mucosal nerves are unaffected by either division of the spinal nerves or mesenteric nerve crushes (Pitha 1969) and are therefore presumed to be intrinsic. Transplantation and lesion studies in pigs have also suggested that the majority of mucosal nerve fibres are of intrinsic origin (Malmfors et al. 1981).

It has usually been assumed that the majority of intrinsic mucosal nerve fibres arise from cell bodies in submucous ganglia, and in many cases they have been traced to this source (Remak 1858; Reichert 1859; Krause 1861; Breiter and Frey 1862; Goniaew 1875; Drasch 1881; Müller 1892; Berkley 1893; Dogiel 1896; Kuntz 1922; Hill 1927; Okamura 1929; Oshima 1929; Ohkubo 1936; Honjin 1951; Temesrékási 1955; Walter 1956; Rintoul 1960; Schofield 1960; Stach and Hung 1979; Furness et al. 1985; Maeda et al. 1985). The interrelationship between submucosal and mucosal nerve fibres is so close in the small intestine that Stöhr (1952) considered the nerves could be regarded as one plexus. Nevertheless, some mucosal nerve fibres arise from myenteric ganglia, as has been observed directly by Dogiel (1896) and Furness et al. (1985).

Recent lesion studies have defined the origins of mucosal nerve fibres in the guinea-pig small intestine, in particular those which contain peptides (Furness and Costa 1987). The types of lesions used in these analyses are illustrated in Fig. 2. Myectomy operations, in which the longitudinal muscle and

Fig. 2a,b. Operations used to lesion intrinsic nerve pathways in the guinea-pig small intestine. The *heavy lines* indicate the sites of lesions and the *dashed* processes of neurons show those parts, which, being severed, would degenerate after the operation. **a** Myectomy. The longitudinal muscle and myenteric plexus is removed from a length (5 – 10 mm) of intestine. Any nerve fibres remaining in the underlying mucosa must have arisen from cell bodies located in the submucous plexus, or in pathways running in the submucosa from more oral or anal ganglia, or from extrinsic ganglia. **b** Homotopic autotransplant. Any remaining fibres in the mucosa must arise from local enteric ganglia or extrinsic nerves. Any fibres that persist in the mucosa after myectomy, homotopic autotransplant and extrinsic denervation must arise from local submucous ganglia; *mp* myenteric plexus, *smp* submucous plexus; *m* mucosa

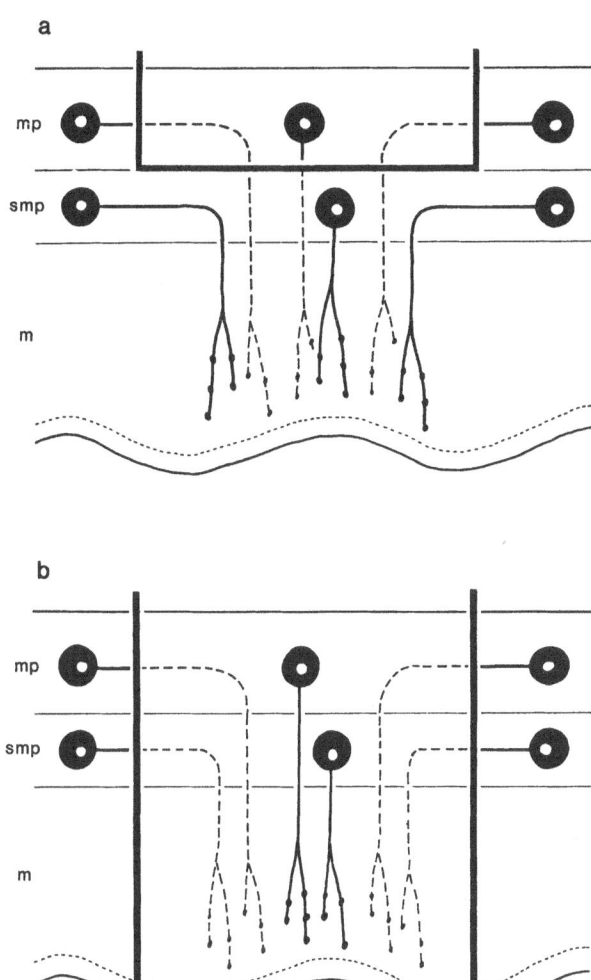

myenteric plexus is removed from a small length of gut, have shown that, with the exception of some substance P (SP) fibres, all of the mucosal nerve fibres appear to remain after operation and are therefore likely to arise from submucous rather than myenteric ganglia. Homotopic autotransplants, in which the wall of the intestine is completely severed at two sites and rejoined in its original position, showed that in sections cut close (2 – 4 mm) to the sites of lesion, no significant differences were seen in the density of peptide-containing mucosal nerve fibres (Keast et al. 1984b). Taken together, these experiments have shown that the majority of mucosal nerve fibres in this region arise from submucous ganglia and travel for only short distances (i.e. less than a few mm) along the gut before projecting to the mucosa. Nevertheless, it should be realized that the degeneration of a small number of fibres after operations may not necessarily be detected, and that other projection patterns may exist for a small proportion of mucosal nerve fibres. In dog small intestine it also seems that the majority of mucosal nerve fibres originate from submucous ganglia (Daniel et al. 1987).

Nerve cell bodies are occasionally found in the mucosa of the small intestine (Drasch 1881; Ishikawa 1926; Stöhr 1934; Ito and Kubo 1940; Newson et al. 1979a; Schultzberg et al. 1980; Dahlström et al. 1984; Keast et al. 1985b) and seem to be more common in the crypt region. The processes of these nerve cells ramify in the small surrounding area of lamina propria. Mucosal nerve cell bodies are more common in the large intestine, again particularly in the middle and basal glandular regions (Reiser 1933; Ohkubo 1937; Ito and Kubo 1940; Lassmann 1975). Stöhr (1934) suggested that mucosal nerve cells are "ectopic" submucous ganglion cells, having similar morphological characteristics to submucous neurons. Nevertheless, some inaccurate identification of neuronal perikarya has arisen due to the presence of interstitial cells in the lamina propria. First described in the intestine by Ramón y Cajal (1894; 1911), these stellate cells, with many long, interweaving processes, have some morphological similarity to neurons as well as a reasonably strong affinity for both methylene blue and silver stains. It was considered for some time that they are either small anastomosing ganglion cells or that, having many intimate connections with neuronal processes, they may regulate nerve activity. Recent ultrastructural studies by Desaki et al. (1984) and others suggest that in the mucosa these may correspond to cells which are similar to fibroblasts and form a discontinuous sheath over mucosal capillaries, and possibly have a role in controlling nutrient and water uptake. Their actual physiological role is still unknown.

4 Localization of Mucosal Nerve Fibre Types

4.1 Acetylcholine Nerve Fibres

Recently antisera raised against acetylcholine (ACh) have been developed; however, as yet, there have been no reports of ACh localization in the intestinal mucosa using this method.

Many studies have been carried out using acetylcholinesterase localization as a marker for cholinergic neurons; however, it is now known that many non-cholinergic tissues also contain this enzyme. Ultrastructural studies have sometimes suggested that nerve fibres containing small, clear vesicles represent cholinergic fibres; however, this definition is also questionable (Gibbins 1982). The best technique available to date is the immunohistochemical localization of the synthesizing enzyme for acetylcholine, choline acetyltransferase (ChAT). Using such antibodies specific populations of enteric neurons (assumed to be cholinergic) have been identified at the light microscope level in both myenteric and submucous ganglia of the guinea-pig small intestine (Furness et al. 1983a, 1984). From the detection of peptides in many of these cholinergic submucosal neurons it is known that they send processes to the mucosa, even though these processes are not usually well-stained by ChAT antisera. The distribution of peptides in cholinergic neurons is discussed below. From studies tracing nerve pathways or observations of nerve populations after specific nerve lesions, it is clear that a small number of extrinsic (vagal) nerve fibres also supply the intestinal mucosa (see above). These fibres have not yet been identified directly by histochemical means.

4.2 Noradrenaline Nerve Fibres

Many descriptions of the noradrenergic innervation of the intestinal mucosa have been published (Norberg 1964; Jacobowitz 1965; Baumgarten 1967; Gabella and Costa 1967, 1968; Pick 1967; Read and Burnstock 1968a, b; Costa and Gabella 1971; Silva et al. 1971; Krokhina 1973; van Driel and Drukker 1973; Schultzberg et al. 1980; Llewellyn-Smith et al. 1984b). These fibres have been identified by fluorescence histochemical or immunohisto-chemical methods (e.g. using antibodies against dopamine-β-hydroxylase), or by the chromaffin reaction. Noradrenergic nerve fibres comprise only a small percentage of the total population of mucosal nerve fibres, are usually found singly and are frequently associated with mucosal blood vessels. They are more prevalent surrounding the intestinal glands, although fine nerve fibres may be found in the villi. The nerve fibres within the villi are usually found in the cores of the lamina propria, sometimes associated with the central lacteal (Thomas and Templeton 1981). Subepithelial nerve fibres are found around the crypts, but are rare in the villi. In guinea-pig small intestine some of the mucosal noradrenergic nerve fibres also contain somatostatin (Costa and Furness 1984).

In the small intestine, the mucosal noradrenergic nerve fibres arise from extrinsic ganglia, as all fibres disappear 2–8 days after extrinsic denervation (Silva et al. 1971; Keast et al. 1984b) or autotransplantation (Malmfors et al. 1981). There are no intrinsic noradrenergic neurons in the small intestine of guinea-pigs, although there are a small number in the proximal colon myenteric plexus (Costa et al. 1971). It is likely that these neurons do not project to the colonic mucosa, as all mucosal nerve fibres disappeared after extrinsic denervation (Mazzanti et al. 1972; Gabella and Juorio 1975). In addition there is a small population of mucosal nerve fibres which are capable of taking up amines and decarboxylating aromatic amino acids (Furness and Costa 1978); it is not known which other substances these neurons contain and use as a transmitter.

4.3 Peptide-Containing Nerve Fibres

Many populations of peptide-containing enteric neurons have been identified using immunohistochemical techniques (see reviews by Schultzberg et al. 1980; Sundler et al. 1980, 1982; Furness et al. 1986). The best-characterized of these nerve fibre types are those containing vasoactive intestinal peptide (VIP), substance P (SP), somatostatin (SOM), neuropeptide Y (NPY) or enkephalin (ENK). Each of these peptides is found in most mammalian species, although a comparison of mucosal innervation between guinea-pigs, rats, dogs, marmosets, rabbits and humans has shown that the innervation density and the proportional representation of each peptide varies considerably along the intestine and between species (Keast et al. 1985 b, 1987). A more detailed discussion of each of these major nerve fibre populations in the mucosa follows. Here the immunoreactivity will be referred to as due to the peptide it is thought to represent (by the results of absorption tests), although further characterization studies have not been carried out in each case.

4.3.1 Vasoactive Intestinal Peptide

VIP has been measured by radioimmunoassay (RIA) in isolated mucosa of the small intestine (Bryant et al. 1976; Dimaline and Dockray 1978; Gaginella et al. 1978a; Furness et al. 1980; Yanaihara et al. 1980; Ferri et al. 1983). Purification of mucosal VIP using high-pressure liquid chromatography (HPLC) has shown that, along with authentic VIP, other forms of VIP exist (Dimaline and Dockray 1978), including a larger form which may not be present in muscle (Yanaihara et al. 1980).

In all species examined so far, immunohistochemical studies have shown that mucosal VIP fibres form a dense network around the crypts and in the villi; many fibres run close to the epithelium, while others are associated with mucosal blood vessels (Larsson et al. 1976; Jessen et al. 1980; Schultzberg et al. 1980; Reinecke et al. 1981; Ferri et al. 1982, 1983, 1984; Fehér and Léránth 1983; Tange 1983; Daniel et al. 1985; Keast et al. 1985 b, 1987). In ultrastructural studies VIP immunoreactivity has been localized to large granular vesicles in mucosal nerve profiles (Larsson 1977; Fehér and Léránth 1983); in human small intestine and guinea-pig colon many nerve endings are seen in the mucosa, particularly next to blood vessels. VIP nerve cell bodies have occasionally been found in the intestinal mucosa of rats, guinea-pigs, dogs and humans (Schultzberg et al. 1980; Keast et al. 1985 b). Immunohistochemical studies using well-characterized antibodies have shown that, in mammals, VIP is not found in any intestinal epithelial cells (Larsson et al. 1979; Dimaline et al. 1980). Recently it has been reported that PHI (peptide histidine isoleucine) is part of the precursor for VIP and coexists with it in many intestinal (including mucosal) nerve fibres (Yanaihara et al. 1983; Ekblad et al. 1985).

Mucosal VIP innervation arises primarily from enteric ganglia, as surgical vagotomy (rabbits), chemical sympathectomy (mice), mesenteric nerve crushes (guinea-pigs), autotransplantation or vagal denervation (pigs) cause no detectable loss of VIP fibres (Larsson et al. 1976; Malmfors et al. 1981; Costa and Furness 1983). In guinea-pigs and dogs it is likely that the majority of mucosal VIP nerve fibres arise from local submucous ganglia, as their distribution is unaffected at the sites of myectomy or myotomy (Costa and Furness 1983; Daniel et al. 1986), or close to homotopic autotransplants (Keast et al. 1984b). In rat small intestine a VIP nerve fibre has been traced directly to the mucosa from a submucous neuron (Maeda et al. 1985). Lesion studies of this type have not been done in other species; however, it is possible that in all mammalian species mucosal VIP innervation arises from submucous ganglia as VIP nerve cells are always found in this layer (Fuxe et al. 1977; Schultzberg et al. 1980; Reinecke et al. 1981; Keast et al. 1985b). Moreover, in Hirschsprung's disease the degree of submucous ganglion cell loss is closely correlated with the loss of mucosal VIP (and SP) nerve fibres (Taguchi et al. 1983; Tsuto et al. 1983; Kishimoto et al. 1984). It is interesting that in the aganglionic segment a few mucosal VIP (but no SP) nerves remain, which the authors suggest may arise from the ganglionated segment via long projections, but may come from extrinsic ganglia. Whether such long pathways also exist in the normal state is not known.

4.3.2 Substance P

Authentic SP has been detected by RIA in the mucosa of the intestine of guinea-pigs, rabbits, rats and humans (Holzer et al. 1982; Ferri et al. 1983; Llewellyn-Smith et al. 1984a). Immunohistochemical studies have shown that SP is primarily located in nerve fibres, although in some species immunoreactive endocrine cells are also seen (Nilsson et al. 1975; Pearse and Polak 1975; Heitz et al. 1976; Ferri et al. 1983; Keast et al. 1985b, 1987).

SP nerve fibres are found throughout the lamina propria of rat, guinea-pig, dog and human intestine (Hökfelt et al. 1975; Schultzberg et al. 1980; Ferri et al. 1982, 1983, 1984; Brodin et al. 1983; Tange 1983; Llewellyn-Smith et al. 1984a; Lolova et al. 1984; Matthews and Cuello 1984; Daniel et al. 1985; Keast et al. 1985b), where some fibres run close to the epithelium or with small blood vessels. SP fibres are prevalent in these species, although they are usually slightly outnumbered by VIP nerve fibres. However, in some other species there are far fewer SP nerve fibres in the intestinal mucosa. In the pig small intestine mucosal SP fibres are absent (Malmfors et al. 1981). In addition Brodin et al. (1983) could not find any in the intestinal mucosa of cats, although Fehér and Wenger (1981) did find SP nerve fibres in both crypt and villus regions. SP fibres are extremely rare in the intestinal mucosa of rabbits (Keast et al. 1987). In human small intestine SP nerve fibres contained both small and large round vesicles (Llewellyn-Smith et al. 1984a), whereas in cats such fibres usually contained large granular vesicles (Fehér and Wenger 1981).

The majority of mucosal SP nerve fibres are derived from enteric ganglia, as shown by severing inputs from vagal or sympathetic ganglia, with no subsequent changes in mucosal SP innervation (Costa et al. 1981; Matthews and Cuello 1984). Capsaicin treatment, which is known to destroy the terminals of primary afferent neurons, has no noticeable effect on the pattern of mucosal SP innervation in guinea-pig (M. Costa and J. B. Furness, unpublished observations) or rat (Hoyes and Barber 1981) small intestine. It is possible, however, that there is a small population of extrinsic SP fibres, the disappearance of which would not be readily detected. In guinea-pig ileum, the majority of the mucosal SP nerve fibres come from local submucous ganglia (Keast et al. 1984b), while a small number arise from the underlying myenteric plexus (Costa et al. 1981).

4.3.3 Somatostatin

Somatostatin has been detected by RIA in the mucosa of the small and large intestine (Furness et al. 1980; Trent and Weir 1981; Vinik et al. 1981; Ferri et al. 1983; Penman et al. 1983; Keast et al. 1984a; Baldissera et al. 1985). Most of this immunoreactivity has been commonly attributed to endocrine cells, which contain a high concentration of SOM and stain brightly in immunohistochemical studies, and little attention has generally been paid to a possible neural source of mucosal SOM. However, Hökfelt et al. described mucosal SOM nerve fibres in rats as early as 1975 and they have since been found in guinea-pigs, cats, dogs, marmosets, rabbits and humans (Costa et al. 1980; Schultzberg et al. 1980; Tange 1983; Lolova et al. 1984; Daniel et al. 1985; Keast et al. 1985b, 1987). A number of studies have suggested that the molecular weight of SOM found in endocrine cells is larger than that in nerves (Trent and Weir 1981; Vinik et al. 1981; Ito et al. 1982; Penman et al. 1983; Baskin and Ensinck 1984; Baldissera et al. 1985). It is possible that many antisera used for immunohistochemistry preferentially recognize this larger molecule. Use of such antisera may form a partial explanation for the common, but mistaken, observation by many authors that SOM nerves were absent from human intestine, whereas SOM endocrine cells were prevalent.

Mucosal SOM nerves form a sparse network, primarily around and below the crypts, which is usually far less prominent than those of SP or VIP nerve fibres in the same region. In guinea-pig ileum, with the exception of a small number of noradrenergic nerve terminals which also contain SOM (Costa and Furness 1984), all of the mucosal SOM nerve fibres arise from enteric ganglia (Costa et al. 1980). Most come from local submucous ganglia, but a small number come from myenteric ganglia (Keast et al. 1984b; Furness et al. 1985).

As yet, to my knowledge, there have been no ultrastructural analyses of mucosal SOM nerve fibres.

4.3.4 Neuropeptide Y

NPY has been found in specific populations of noradrenergic neurons throughout the body, including some supplying the gastrointestinal tract. However, in the intestine there are many more intrinsic NPY nerve fibres which do not contain noradrenaline (NA; Furness et al. 1983b, 1985; Sundler et al. 1983).

Mucosal nerve fibres containing NPY have been localized using antisera directed either towards this peptide or to a similar peptide, pancreatic polypeptide (PP). Although there is considerable structural homology between NPY and PP, it is now thought that NPY is found exclusively in nerves, whereas PP and the related peptide, PYY (peptide YY), are found in endocrine cells (Lundberg et al. 1982; Emson and de Quidt 1984). NPY nerve fibres have been found in the intestinal mucosa of guinea-pigs, rats, mice, cats, dogs, marmosets, rabbits and humans (Lorén et al. 1979; Furness et al. 1983b; Sundler et al. 1983; Taylor and Vaillant 1983; Daniel et al. 1985; Keast et al. 1985b, 1987). Many submucosal nerve fibres, along with some in the mucosa, are associated with blood vessels (and probably of extrinsic origin), whereas the remaining mucosal NPY nerves form quite a dense irregular network throughout the lamina propria. Some fibres lie close to the epithelium of the crypts and villi.

In rats, the mucosal innervation was unaffected by either vagotomy or upper abdominal sympathectomy (Sundler et al. 1983); similarly, surgical or chemical sympathectomy had no effect on mucosal NPY nerve fibres in the guinea-pig (Furness et al. 1983b). These studies suggest that most of the mucosal NPY nerves arise from enteric ganglia. Other studies in the guinea-pig have shown that the majority of mucosal NPY nerves arise from local submucous ganglia (Keast et al. 1984b), while a much smaller number, which also contain other peptides (see below), come from myenteric ganglia (Furness et al. 1985).

4.3.5 Enkephalin

Nerve fibres containing leu- or met-enkephalin have been detected in the gastrointestinal tract, but are very scarce or absent in the intestinal mucosa of all species so far examined (Schultzberg et al. 1980; Ferri et al. 1982, 1984; Furness et al. 1983c; Lolova et al. 1984; Tange 1983; Kobayashi et al. 1984; Keast et al. 1985b, 1987). As there are not known to be any differences in the mucosal distribution of these two opiates, they will be referred to collectively as enkephalin (ENK) fibres. In the intestine ENK-immunoreactivity (IR) is found in a very small population of mucosal nerve fibres, located almost exclusively in the muscularis mucosae and around the bases of the glands. Very few fibres have been observed in the villi or superficial lamina propria of the small or large intestine. Nerve fibres containing dynorphin (DYN)-immunoreactivity have been found in the mucosa of the colon, and a small number in the duodenum of the guinea-pig; in the former tissue, immunoreactive submucous neurons have also been found, which may be the source of the mucosal nerve fibres (Vincent et al. 1984).

The origin of the ENK-IR mucosal nerve fibres is unknown. There are no ENK neurons in the submucous ganglia of rat or guinea-pig small intestine (Schultzberg et al. 1980; Furness et al. 1983c), implying that fibres must arise from either myenteric or extrinsic ganglia.

4.3.6 Other Neuropeptides

Many other peptides have been discovered in enteric nerve fibres, including those in the mucosa. These include gastrin-releasing peptide (GRP; Dockray et al. 1979; Moghimzadeh et al. 1983; Costa et al. 1984; Leander et al. 1984; Keast et al. 1987), cholecystokinin (CCK; Schultzberg et al. 1980; Leander et al. 1984), galanin (Melander et al. 1985), and calcitonin gene-related peptide (CGRP; Clague et al. 1985; Furness et al. 1985). As these nerve fibres have not yet been described in many areas and species, they will not be mentioned further in this general summary.

4.4 Peptide Coexistence

In the past few years immunohistochemical techniques have been developed which allow simultaneous localization of two antigens within the one preparation. The enteric nervous system of the guinea-pig has been extensively studied in this way and the majority of neurons have been found to contain more than one peptide (Fig. 3). Some of these neurons also contain ChAT and are assumed to be cholinergic. In the submucous ganglia, which are the source of the majority of peptide-containing mucosal nerve fibres (see Sect. 3), approximately 45% of all neurons contain both VIP and DYN, while the remaining approximately 55% contain ChAT. These cholinergic neurons can be classified on the basis of the peptides they contain (Furness et al. 1984, 1985); approx. 29% contain CCK, CGRP, NPY and SOM, another approx. 11% contain SP, while the remaining cholinergic neurons do not contain any of the peptides studied so far. As most of the peptide-containing nerve fibres in the mucosa arise from submucous ganglia, the patterns of coexistence observed for the submucous neurons could be expected to apply to their projections to the mucosa. It is possible, however, that the relative amounts of each peptide

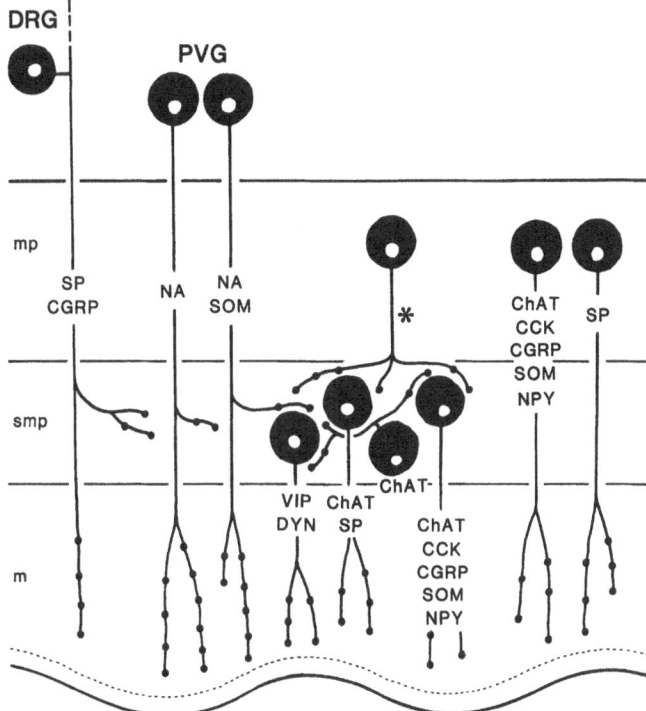

Fig. 3. Chemical coding and origins of nerve fibres in the mucosa and submucosa of the guinea-pig small intestine. Nerve fibres in the mucosa arise from extrinsic and intrinsic ganglia. Sensory nerve fibres containing both *SP* and *CGRP* arise from dorsal root ganglia (*DRG*). Sympathetic noradrenergic neurons from prevertebral ganglia (*PVG*) also travel to the mucosa and some contain *SOM*. The majority of mucosal nerve fibres come from local submucous ganglia. Some of these contain both *VIP* and *DYN*, while the remaining submucous neurons contain ChAT and are assumed to be cholinergic. Some of these cholinergic neurons contain *SP*, some contain *CCK, CGRP, NPY* and *SOM*, while others contain none of these peptides. It is not known whether this latter group of neurons project to the mucosa, although they are deduced to supply submucous neurons (see text). Two groups of myenteric neurons also project to the mucosa, *SP* neurons and *ChAT* neurons which contain *CCK, CGRP, NPY* and *SOM*. Inputs to submucous ganglia come from extrinsic ganglia (*DRG* and *PVG*) and myenteric neurons. Several types of myenteric neurons (indicated by *) project to submucous ganglia; substances in these neurons include *CCK, DYN, ENK, GRP, 5-HT, SOM, SP* and *VIP*

may differ in the processes (compared with the cell body); this is exemplified by CCK and SOM, which coexist in submucous nerve cell bodies, and yet in the mucosa there are many more nerve terminals which show immunoreactivity for SOM than for CCK.

It is also known that a small number of myenteric neurons project to the mucosa (see Sect. 3). Some of these contain the combination ChAT/CCK/CGRP/NPY/SOM (as seen in some submucous neurons); it is not known whether the SP neurons which project from the myenteric plexus to the mucosa also contain ChAT.

5 Mucosal Functions and Intestinal Reflexes

It is concluded that there are sensory nerve fibres in the mucosa, as some chemical and mechanical stimuli applied to the mucosa can elicit changes in intestinal motor, vascular and secretory activity (e.g. Bülbring et al. 1958; Hukuhara et al. 1958; Biber et al. 1971). Many of these physiological experiments suggest that both intrinsic and extrinsic reflex pathways modify mucosal function, implying that there are also motor fibres to the epithelium (Müller 1911; Brunemeier and Carlson 1914; Ranson 1921; Schofield 1960; Caren et al. 1974). Because most of the mucosal nerve fibres arise from submucous ganglia, these ganglia are likely to be important sites for regulating mucosal function. Preganglionic nerve terminals on submucous neurons, arise from vagal, sympathetic and myenteric and submucous neurons, and submucous neurons may also send processes to the myenteric plexus (Ramón y Cajal 1911; Cavazzana and Borsetto 1948). Thus, pathways exist that could enable coordination of mucosal function with other intestinal (e.g. peristalsis, gastric emptying) or extra-intestinal (e.g. pancreatic, biliary, central) events.

Of the many mucosal functions potentially regulated by neural activity, only transepithelial water and electrolyte movement will be discussed further. However, it should be borne in mind that other factors (e.g. blood flow, endocrine and mucous secretions) are likely to influence the net state of mucosal transport. "Secretion" is used to describe net movement of a substance in the lumen, whereas "absorption" refers to movement from the lumen to the interstitial fluid.

The primary function of the mammalian small intestine is the digestion and absorption of nutrients and water. In the small intestine the epithelium is highly permeable to water and electrolytes. Large bidirectional fluxes of water, sodium and chloride pass across the epithelium, but most of this transport occurs passively, by a paracellular route. Some net secretion of bicarbonate may also occur. Net absorption of water and ions is usually observed, and this is ultimately dependent on active transcellular transport processes (primarily the electrogenic sodium pump on the basolateral membrane of the epithelial cell), which are responsible for setting up ionic and osmotic gradients across the epithelium. These gradients therefore provide the driving forces for ion uptake, water then following passively. During exposure to substances such as cholera toxin (CT), bile salts, theophylline and prostaglandins (PGs), active chloride secretion is stimulated; the physiological role of active secretion is not known, although some possibilities are discussed below. There is spatial separation of active absorption and secretion, with secretion occurring primarily in the crypts, and absorptive processes mainly localized to the villi; this reflects differences in the nature of the epithelial cell layer in these two regions.

A similar absorptive function is provided by the large intestine, although the epithelium is less permeable to passive water and ion fluxes, so that larger concentration gradients can be maintained and the luminal content derived from the small intestine concentrated further. In addition potassium and bicarbonate secretion occur.

6 Stimulation of Mucosal Nerve Fibres In Vitro

It is possible to stimulate mucosal nerve fibres electrically in vitro, in a modified Ussing chamber (Ussing and Zerahn 1951), and to monitor effects on transport with an automatic voltage clamp. In the standard method a small piece of intestine, usually with external muscle layers removed, is mounted in a perfusion chamber. The potential difference (PD) generated across the tissue is clamped at zero throughout the experiment; the current required to offset the tissue PD is conventionally referred to as the short-circuit current (I_{sc}) and, along with the measurement of transmembrane conductance (G), gives a good indication of net active ion transport and permeability of the tissue. Radioisotope substitutions of permeant ions can be used to define the ionic basis of any changes in I_{sc}. Where these details are not described below, these analyses have not been carried out. By passing a current between electrodes placed close to the tissue, the nerves within it can be stimulated, a procedure referred to as electrical field stimulation (EFS).

EFS causes a marked transient rise in I_{sc} in mucosa-submucosa preparations isolated from the small intestine of rabbits (Hubel 1978; Hubel and Callanan 1980), humans (Hubel and Shirazi 1982) and guinea-pigs (Cooke et al. 1983 b; Keast et al. 1985 c), and the human (Hubel et al. 1983) and canine (Rangachari and McWade 1986) colon. These I_{sc} increases are primarily due to stimulation of chloride secretion, require extracellular calcium in the serosal bathing fluid, and are abolished by tetrodotoxin (TTX), veratridine, high potassium levels or scorpion venom (which initially depolarizes nerves and eventually blocks transmission). The response to EFS in rabbits and guinea-pigs was mimicked quite closely by the initial (depolarizing) response to scorpion venom (Cooke et al. 1983 a; Hubel 1983). The studies of Carey et al. (1985) and Rangachari and McWade (1986) suggest that, in guinea-pigs and dogs, respectively, EFS has no measureable direct effects on active ion transport. Taken together, these studies indicate that there are secretomotor nerve fibres in these tissues which can be stimulated electrically. In most of these experiments only the mucosa and submucosa are present, so it is likely that either submucous neurons or fibres in the mucosa are stimulated by this technique.

The relative sizes of the cholinergic and non-cholinergic components of the secretory response to EFS differ between species. In guinea-pig and human intestine the responses to EFS were substantially reduced by atropine or hyoscine; however, in rabbit ileum they were only slightly reduced by atropine, even though there was a more substantial cholinergic component of the secretory response to scorpion venom in this species (Hubel 1978, 1983).

It is not known which neurotransmitter(s) are responsible for the hyoscine-resistant responses to EFS, although in the rabbit ileum these responses are unaffected by adrenoreceptor antagonists, hexamethonium, somatostatin and desensitization to substance P (Hubel 1984). In the guinea-pig the reponse to EFS is unchanged by the 5-HT antagonist cisapride or 5-HT desensitization (Cooke and Carey 1985). Recent studies using a SP analogue which can reduce the effect of SP on mucosal epithelial cells, but has little or no effect on neuronal SP receptors, suggest that some of the secretion caused by EFS is due to the stimulation of submucous neurons which release SP (Keast et al. 1985 c).

A likely contributor to the remaining atropine-resistant responses to EFS in all of the species that have been studied is VIP, which is found in a large number of mucosal nerve fibres in all mammalian species examined so far (see Sect. 4.3.1). VIP is also released by rabbit ileum during EFS (Gaginella et al. 1981), and is a potent stimulant of water and ion secretion, wherever it has been tested (see Sect. 7.3.1). However, it is possible that the release of other substances present in mucosal nerves also contributes to the non-cholinergic response to EFS, as substance P does in the guinea-pig.

Secretomotor neurons in the submucosa can also be stimulated in other ways. In rat colon adenosine 5'-triphosphate (ATP) selectively stimulates non-cholinergic secretomotor neurons (Cuthbert and Hickman 1985). In guinea-pig ileum, where separate populations of cholinergic and non-cholinergic submucous neurons have been demonstrated immunohistochemically (Furness et al. 1984); these two groups can also be stimulated selectively (Keast et al. 1985 c). We have found that DMPP (1,1-dimethyl-4-phenylpiperazinium) preferentially stimulates cholinergic secretomotor neurons, 5-hydroxytryptamine (5-HT or serotonin) in a concentration of approximately 10^{-7} M preferentially stimulates non-cholinergic neurons (i.e. the VIP neurons, in this species), and higher concentrations of 5-HT and EFS stimulate both neuronal populations (Figs. 4, 5). There is no evidence as yet for the existence of submucous neurons which, when stimulated, cause an increase in net absorption. Thus, absorption cannot be enhanced directly by nerves but can be enhanced by reducing the activity of these secretomotor neurons.

In rabbit ileum and rat and dog colon TTX causes a decrease in I_{sc}, representing an increase in sodium and chloride absorption and a decrease in residual ion fluxes (probably representing a decrease in HCO_3^- secretion; Hubel 1978; Andres et al. 1985; Cuthbert and Hickman 1985; Rangachari and

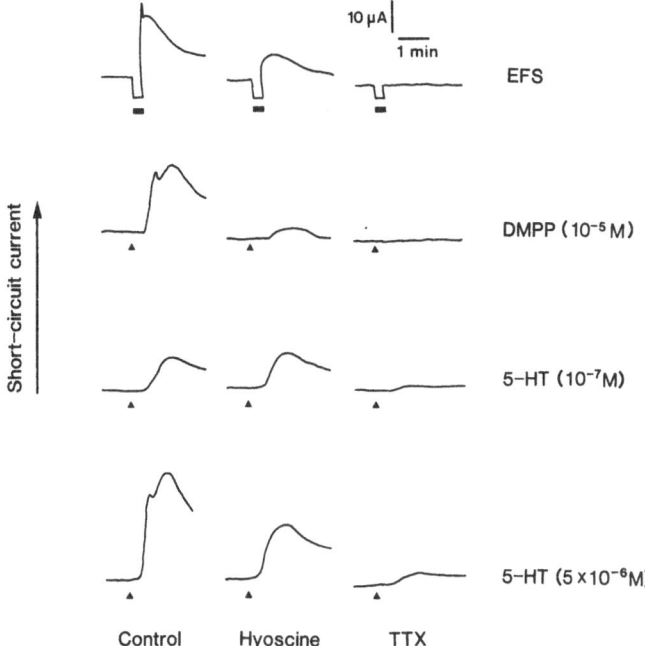

Fig. 4. Effects of EFS, DMPP and 5-HT on mucosal transport, measured in terms of short-circuit current (I_{sc}) of guinea-pig ileum mucosa-submucosa. Representative records of I_{sc} from four tissues are shown. For each tissue, the effects of hyoscine and TTX (each $10^{-7} M$) are indicated. In the *top trace* the period of EFS is indicated by the *bar*; bipolar rectangular current pulses of 20 V peak-to-peak amplitude, 10 Hz, 0.5 ms duration were passed across the tissue for 20 s periods. Except for the period of EFS the PD was clamped at zero throughout all experiments. Times of DMPP and 5-HT addition (both to the submucosal bathing solution) are indicated by *arrowheads*. Between each period of EFS or drug addition, at least 10 min were allowed. DMPP and 5-HT were washed out shortly after the maximum I_{sc} had been obtained (washout period not shown here). Hyoscine and TTX were present for at least 10 min in both mucosal and submucosal bathing solutions, prior to EFS, DMPP or 5-HT. EFS, DMPP and 5-HT all caused an increase in I_{sc}, which was substantially reduced by TTX, indicating nerve-dependent responses. The majority of the response to DMPP was also reduced by hyoscine, indicating a predominant action on cholinergic neurons, whereas the responses to EFS or 5-HT ($5 \times 10^{-6} M$) were reduced by hyoscine to a lesser extent. These then act on both cholinergic and non-cholinergic neurons. Lower concentrations of 5-HT (e.g. $10^{-7} M$) caused an increase in I_{sc} which was only partly affected by hyoscine, indicating that the response was dependent mainly on non-cholinergic neurons. (Details of experiments given in Keast et al. 1985c)

McWade 1986); in the colon of rats and dogs this effect of TTX is not seen if the submucosa is removed, implying that in vitro, too, intrinsic secretomotor neurons are tonically active. In these tissues neuronal secretomotor activity causes fluctuations in baseline I_{sc}, which are abolished by TTX. The existence of continuously active secretomotor nerves is less certain in isolated guinea-pig ileum. Cooke (1986) has reported that TTX causes a significant decrease in I_{sc} only when glucose is absent from the mucosal bathing solution, whereas in our experiments TTX causes a signifcant decrease in I_{sc}

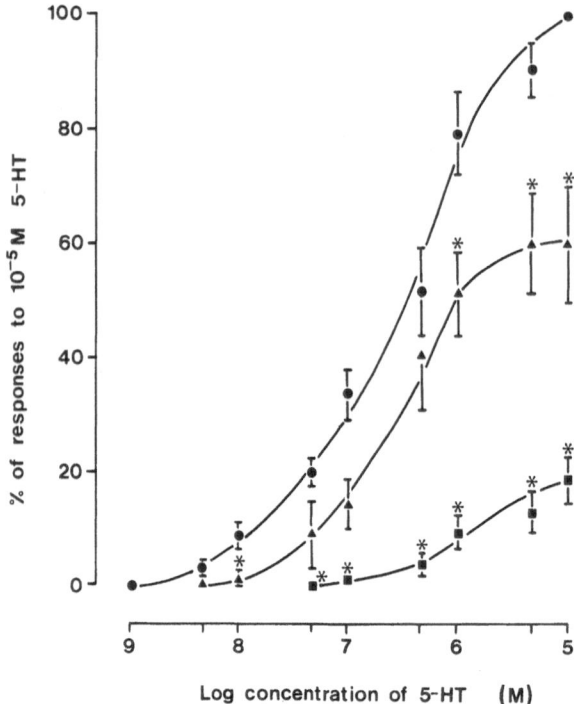

Fig. 5. Concentration-response curves for increases in I_{sc} elicited by 5-HT in guinea-pig ileum mucosa-submucosa. 5-HT was added to the submucosal bathing solution in increasing concentrations, with a washout period between each addition. For each tissue, two such curves were generated, one a control (●) and the other in the presence of either $10^{-7}\,M$ hyoscine (▲) or $10^{-7}\,M$ TTX (■). The responses to $10^{-5}\,M$ 5-HT in the absence of hyoscine or TTX were taken as the response maxima, and all values are expressed as a percentage of these. Each point represents the mean ± SEM for control ($n = 11$), hyoscine ($n = 6$) and TTX ($n = 5$) treated tissues. *Asterisks* represent responses which are significantly different from control responses to the same concentration of 5-HT in the absence of antagonist ($P < 0.05$, paired two-tailed t-test). These results indicate that a major component of the response to 5-HT is dependent on nerve activity and that hyoscine has a significant inhibitory effect on responses to 5-HT, particularly when the concentration of 5-HT is $\geq 10^{-6}\,M$. (Reprinted from Keast et al. (1985c) with minor modifications, with permission)

whether glucose is present or absent (unpublished observations). The reason for this discrepancy is not known, but may be related to differences in tissue preparation and handling.

In guinea-pigs and rabbits secretomotor neurons which are continuously active in vitro are probably non-cholinergic, as exposure to hyoscine has negligible effects on I_{sc}. In contrast, in in vivo studies in cats and dogs, atropine causes an increase in absorptive fluxes, suggesting ongoing activity in cholinergic neurons.

7 Effects of Substances Found in Mucosal Nerve Fibres on Epithelial Transport

A summary of the effects on mucosal transport of the substances found in each major group of mucosal nerve fibres follows and is shown in Table 1. Each substance is described in terms of its effects on the small and large intestine; major differences in results between in vivo and in vitro experiments are also mentioned. Absorption and secretion generally refer to the movement of both water and ions (usually only sodium and chloride are monitored), although in some cases only water or only ion fluxes were recorded.

The most common way of investigating the effects of exogenous substances on mucosal transport in vivo involves cannulation of a small length of intestine; after administration of drugs (by either intravenous, intra-arterial or intra-luminal routes) the water and ion uptake by the segment is determined (commonly by the use of radioactive non-absorbable solutes and radioisotopes, respectively). In humans a triple lumen perfusion catheter is used for a similar type of investigation. In most of the in vivo studies changes in motility and blood flow are not described or accounted for. The majority of in vitro studies utilize Ussing chambers or isolated sac preparations and,

Table 1. Actions of substances found in mucosal nerve fibres on mucosal transport

Substance	Action[a]	Effect on I_{sc}	Probable site(s) of action	Comments
Cholinomimetics	Secretion	Increase	Epithelium and nerves	Muscarinic receptors on epithelial cells Nicotinic receptors on submucous neurons
VIP	Secretion	Increase	Epithelium	Receptors on epithelial cells Response unaffected by TTX or removal of submucous neurons
Substance P	Secretion	Increase	Epithelium and nerves	
Adrenomimetics	Absorption	Decrease	Epithelium and nerves	Size of effect on epithelial cells variable between species Usually act on a-receptors
Somatostatin	Absorption	Decrease	Epithelium and nerves	Effect on nerves only tested in guinea-pig ileum
Neuropeptide Y	Absorption	Decrease	Epithelium	
Opiates	Absorption	Decrease	Nerves	Probably act on δ-receptors

[a] Refers to net direction of effect on water and ion transport

when results from these studies are taken together, effects on the epithelium alone can be defined.

The limited amount of information to date suggests that there are coordinated fluctuations in transmural PD and muscular contraction (Read et al. 1977; Greenwood and Davison 1985). During these PD changes the lumen became more negative, possibly representing stimulation of anion secretion. In ferrets simultaneous changes in PD and contraction can be elicited by vagal stimulation; basal contractility and PD fluctuations can be reduced by atropine, TTX or vagotomy, but are unchanged by cutting the splanchnic nerves. Vagal activity can therefore alter both muscular and mucosal functions, but as yet coordinated effects of activity in intrinsic neurons have not been studied.

7.1 Cholinomimetics

Small intestine. Cholinomimetic agents cause a net secretion of water, Na and Cl into the lumen of the small intestine, an action which is blocked by atropine or hyoscine, mimicked by eserine and potentiated by neostigmine (Tidball 1961; Hardcastle and Eggenton 1973; Hubel 1976, 1977; Isaacs et al. 1976; Hardcastle et al. 1981 b; Cooke 1984). This secretion is represented in vitro by an increase in I_{sc}.

The secretory effects of muscarinic agonists appear to be primarily on crypt cells (Browning et al. 1978). Cyclic AMP levels in the epithelial cells are unaffected by these agonists (Schwartz et al. 1974; Isaacs et al. 1976; Laburthe et al. 1979), but the secretory effects instead appear to cause, and be dependent on, an enhanced influx of free Ca^{++} into the epithelium or subepithelial tissue (Bolton and Field 1977; Donowitz et al. 1982; Hardcastle et al. 1983).

Binding studies using the irreversible muscarinic receptor antagonist quinuclidinyl benzilate (QNB) have demonstrated muscarinic receptors on crypt and villus epithelial cells of the small intestine (Isaacs et al. 1982). In rats, the epithelial muscarinic receptors are primarily on the basolateral membranes (Gaginella 1984) and have different agonist-binding affinity compared with those found in intestinal smooth muscle (Tien et al. 1985) and in guinea-pig ileum are of the M_2 subtype (Cooke 1986). Receptors for acetylcholine have been demonstrated on cultured crypt cells from human foetal small intestine, which are hyperpolarized by this substance; this response is associated with an increase in potassium conductance of the epithelial cell membrane (Yada and Okada 1984). It is likely that substances such as somatostatin (Guandalini et al. 1980; Keast et al. 1986a) and morphine (Turnberg et al. 1982), which can diminish carbachol secretory responses in vitro, do so by a direct effect on epithelial cells. It should be noted that opiates and somato-statin can also inhibit secretion indirectly (see Sects. 7.3.3 and 7.3.5).

Nicotinic receptors are also present in the isolated mucosa-submucosa. In the rabbit ileum, the nicotinic agonist DMPP initially caused an increase in I_{sc} (thought to be due to an action on nicotinic receptors on submucous neurons), followed by a prolonged decrease in I_{sc}, which could be blocked by either hexamethonium or phentolamine (Tapper and Lewand 1981). High concentrations ($\geq 10^{-4}$ M) of carbachol caused a similar phenomenon (Tapper et al. 1978). The decreases in I_{sc} are due to an increase in sodium and chloride absorption and have been interpreted by the authors to be due to an action on nicotinic receptors to release transmitters from noradrenergic nerve fibres in the mucosa. In guinea-pig ileum DMPP also causes an increase in I_{sc}, which is abolished by TTX and largely inhibited by hexamethonium and hyoscine (Keast et al. 1985c). However, even though a decrease in I_{sc} follows (as for the rabbit), this is unlikely to be due to DMPP action on noradrenergic nerve fibres, as a similar late decrease in I_{sc} was also elicited in segments of small intestine which had been extrinsically denervated (i.e. in which the mesenteric nerves supplying that segment of intestine had been crushed and the damaged extrinsic nerve fibres allowed to degenerate; unpublished obser-vations).

In guinea-pig ileum both noradrenaline (NA) and somatostatin alter basal mucosal transport mainly by acting on nerves; in this tissue the secretory component of the response to DMPP (i.e. the rise in I_{sc}) was significantly reduced by NA or somatostatin, suggesting that both of these substances can reduce the activity of cholinergic secretomotor neurons in submucous ganglia (Keast et al. 1986).

Large intestine. Cholinomimetic agents cause an atropine-sensitive secretion and increase in I_{sc} and chloride secretion, similar to that seen in the small in-testine (Browning et al. 1977). The secretion caused by cholinomimetics is potentiated by neostigmine, but unaffected by hexamethonium or TTX (Zim-merman and Binder 1983). Carbachol also caused an increase in I_{sc} and chloride secretion in a human colonic epithelium tumour cell line, grown as a monolayer and perfused in a modified Ussing chamber (Dharmsathaphorn et al. 1984; Dharmsathaphorn and Pandol 1986). DMPP and high concentra-tions of carbachol (10^{-3} M) may also act on nicotinic receptors on noradrenergic axons in rat colon (Zimmerman and Binder 1983), as has been suggested for rabbit ileum (see above).

The effects on I_{sc} and secretion are dependent on the presence of ex-tracellular calcium (Zimmerman et al. 1982; Dharmsathaphorn and Pandol 1986). Saturable high-affinity receptors for ^{3}H-QNB have been demonstrated in the epithelium of rat large intestine (Rimele et al. 1981; Rimele and Gaginella 1982; Gaginella 1984); the concentration of agonist required to increase I_{sc} was well correlated with the agonist affinity for these muscarinic receptors (Zimmerman and Binder 1982, 1983), implying that transport effects of cholinergic agonists could occur via such receptors. There do not seem to be any differences in affinity between the muscarinic receptors (as defined by this method) of rat jejunum, ileum and colon (Rimele and Gaginella 1982).

7.2 Adrenomimetics

Small intestine. Noradrenaline (NA) and adrenaline (A) enhance the absorption of water, Na and Cl across the intestinal mucosa (Aulsebrook 1965b; Hubel 1976; Brunsson et al. 1979; Parsons et al. 1983; Rao et al. 1984); in vitro a decrease in I_{sc} is observed, and represents an increase in coupled Na-Cl absorption usually combined with an inhibition of chloride and bicarbonate secretion (Field and McColl 1973; Field et al. 1975; Durbin et al. 1982). This response is closely mimicked by acute administration of reserpine (Aulsebrook 1965a) or tyramine (Tapper et al. 1981); the tyramine effects were absent in animals which had been pretreated with 6-hydroxydopamine (6-OHDA), an agent which degenerates noradrenergic axons. Rats pretreated with 6-OHDA also have impaired basal fluid absorption compared with control animals (Chang et al. 1985). In the duodenum NA stimulates bicarbonate ion secretion (Flemström et al. 1982).

NA and α-agonists diminish the secretory effects of cholera toxin (CT; Donowitz and Charney 1979), *E. coli* heat-stable enterotoxin (ST; Ahrens and Zhu 1982b), PGE_1 and dibutyryl-cAMP (Nakaki et al. 1982; Bunce and Spraggs 1983), electrolyte perfusion (Schiller et al. 1985) and VIP (Nakaki et al. 1982; Rao et al. 1984).

In most cases, where a variety of agonists and antagonists have been used, α_2-agonists (e.g. clonidine) were the most effective in increasing absorption (Tanaka and Starke 1979; Chang et al. 1982, 1983) or reducing the responses to a variety of secretory stimuli (Nakaki et al. 1982; Bunce and Spraggs 1983; Doherty and Hancock 1983; Schiller et al. 1985). These effects on artificially evoked secretion were blocked by antagonists such as yohimbine, but were unaffected by prazosin, propranolol or naloxone.

The subtype of α-receptor involved may be partially dependent on the species and intestinal site, with α_1-receptors found in some areas of rat small intestine (Cotterell et al. 1983; Parsons et al. 1983). As yet, the only species in which significant β-receptor effects on transport have been described is the human (Morris and Turnberg 1981). There are a number of reports of propranolol (a β-receptor antagonist) decreasing the secretory effects of CT (Donowitz and Charney 1979) and bile acids (Conley et al. 1976; Coyne et al. 1977; Taub et al. 1977), without having any effect on mannitol-induced secretion, basal electrolyte transport or epithelial cAMP levels. These actions could be ascribed to the general anaesthetic effects of high concentrations of propranolol; this would be consistent with the neural mechanism of action proposed for CT and bile acid-induced secretion (see Sect. 10.1), whereas intraluminal mannitol is known to cause secretion by a purely osmotic mechanism and would thus be unaffected by propranolol.

In rabbit ileum and rat jejunum the effects of adrenomimetic agents in vitro were unaffected by TTX (Dobbins et al. 1980; Parsons et al. 1984). Moreover, a_2-receptors have been localized in the isolated mucosa (Chang et al. 1983; Tsai et al. 1985) and consequently adrenomimetics have been assumed to act directly on epithelial cells of the mucosa. Although this is likely to be true for some tissues, experiments in cats and guinea-pigs strongly suggest that NA causes changes in mucosal transport primarily by an action on enteric neurons. In studies on cats in vivo, the absorptive effects of NA were inhibited by TTX (Sjövall et al. 1983c) and in guinea-pigs in vitro, the decrease in I_{sc} was abolished by TTX (Keast et al. 1986). In the latter case, exogenous NA was found to have an inhibitory action on both cholinergic and non-cholinergic submucous neurons.

The experiments in the guinea-pig have some limitations in that only I_{sc} was measured, and the ionic site of action of TTX was not defined. Moreover, NA and TTX both cause a decrease in I_{sc} and, in other studies, cause a net increase in Na and Cl absorption. However, even when baseline Cl secretion is enhanced by theophylline (which acts directly on epithelial cells), the responses to NA were abolished by TTX (Keast et al. 1986). Thus, any effect of NA on Cl movement which decreases I_{sc} is probably dependent on nerve activity. Ion substitution or radioisotope flux studies are needed to clearly identify any direct action of NA on other ion transport processes (e.g. sodium or bicarbonate fluxes) across the epithelium.

There have been no consistent changes in intracellular cyclic nucleotide or calcium concentrations reported for the a-adrenoreceptor agonists (Field et al. 1975).

Large intestine. When administered in vivo, NA enhances basal absorption and reduces secretion caused by VIP (Rao et al. 1984); in vitro NA causes a decrease in I_{sc}, associated with an increase in Na and Cl absorption (Racusen and Binder 1979; Sellin and de Soignie 1985). Beta-agonists cause similar effects in rabbit distal colon (Smith and McCabe 1986).

In the mucosa of rabbit descending colon NA effects are mediated by β-receptors (Smith and McCabe 1984), whereas in rats they involve both a- and β-receptors, and are unaffected by TTX, naloxone and reserpine (Racusen and Binder 1979). Adrenaline (A) causes a reversal of transepithelial PD in preparations of isolated rabbit crypt epithelium (Krasny and Frizzell 1984). In crypt cells from a human colonic tumour, A has no effect on I_{sc} (Dharmsathaphorn et al. 1984), whereas in a similar preparation of normal human colonic epithelial cells A decreased the basal cAMP levels, as well as antagonizing the increases in cAMP due to VIP stimulation, via an a_2-receptor mechanism (Boige et al. 1984). Thus, there appears to be considerable variability between tissues and species in the site of catecholamine action (i.e. whether it acts on neurons, epithelial cells or both) and the receptor type involved.

7.3 Peptides

The extensive distributions of peptides in nerve fibres of the mucosa and submucosa, along with their variety of effects in the gastrointestinal tract, suggest physiological roles for these substances. The subepithelial distribution of many of these nerve fibres in the small intestine indicates a possible action. on epithelial function, although it should always be considered that other actions (e.g. changing blood flow, mucus secretion or muscle contraction) could affect water and ion movement.

7.3.1 VIP

VIP has been implicated in the regulation of absorption and secretion since the discovery of abnormally high concentrations of VIP in the plasma and tissues of patients with the Werner-Morrison syndrome (pancreatic cholera; Bloom et al. 1973; Said and Faloona 1975; Bryant et al. 1976; Udall et al. 1976). These patients are most commonly found to have pancreatic islet cell carcinoma, pheochromocytoma or ganglioneuroblastoma. Such tumours produce and secrete enormous amounts of VIP, as well as a larger form of VIP (Yamaguchi et al. 1980) and PHI (Bloom et al. 1983). It is not clear which other peptides are produced in excessive amounts, but it is commonly assumed that VIP is the primary factor causing the enhanced secretion in these patients, as it can be reproduced by intravenous VIP infusion and is reversed when infusion stops (Modlin et al. 1978; Krejs and Fordtran 1980). VIP effects on mucosal transport have subsequently been studied more thoroughly than those of any other of the neuropeptides.

Small intestine. The secretory effect of VIP in the small intestine was first demonstrated by Barbezat and Grossman (1971) in the dog and has since been reported many times (Schwartz et al. 1974; Krejs et al. 1978; Wu et al. 1979; Beubler 1980; Camilleri et al. 1981; Mitchenere et al. 1981; Granger et al. 1982). This secretion is represented by an increase in I_{sc}, which is primarily attributed to an enhanced chloride secretion (Schwartz et al. 1974). Active glucose absorption is unaffected by VIP (Coupar 1976). Duodenal bicarbonate ion secretion is enhanced by VIP (Isenberg et al. 1984; Flemström et al. 1985).

Under in vivo experimental conditions VIP also causes atropine-resistant hyperaemia in the cat (Eklund et al. 1979). In addition, Krejs et al. (1978) have shown that VIP causes a reversible increase in protein output from and dilatation of the mucosal capillaries in dog jejunum. Conversely, Mailman (1978) has described a VIP-induced atropine-sensitive decrease in intestinal blood flow in the dog. Unfortunately, most other in vivo studies have not mentioned blood flow changes after VIP, and it is not clear whether any of the VIP effects on mucosal transport in vivo are associated with or dependent upon vascular changes.

VIP-induced secretion in cats in vivo (Cassuto et al. 1983) and in rabbit mucosa in vitro is unaffected by TTX (Binder et al. 1984); VIP-induced secretion in rats in vivo is also unaffected by atropine (Beubler 1980). In the isolated guinea-pig mucosa-submucosa the response to VIP is unaffected by removal of the submucosa (Carey et al. 1985). Together these studies suggest that VIP has a direct action on the epithelium. This is supported by studies on cultured intestinal epithelial cells, in which VIP causes hyperpolarization, primarily via an increase in potassium conductance (Yada and Okada 1984). Consequently, opiate agonists, angiotensin II and NA, which reduce VIP-induced secretion (Coupar 1983; Rao et al. 1984), are likely to do so by an effect on epithelial cells.

VIP receptors have been demonstrated on epithelial cells of rat small intestine (Prieto et al. 1979), where they have been localized to the basolateral membrane (Dharmsathaphorn et al. 1983). Binding of VIP to these receptors appears to require the entire sequence of VIP (Prieto et al. 1979; Couvineau et al. 1984), as synthetic fragments of VIP are, at best, approximately 1% as potent as the complete VIP molecule. There are also considerable species differences in binding properties of VIP receptors, more than would be expected from the small differences in VIP sequences (Couvineau et al. 1984). Binding of VIP to epithelial cell receptors is clearly linked with an increase in intracellular cAMP concentration (Schwartz et al. 1974; Klaeveman et al. 1975; Simon and Kather 1978; Laburthe et al. 1979; Beubler 1980, 1981). This is assumed to result in phosphorylation of one or more membrane proteins to cause a change in transepithelial ion movement.

Large intestine. VIP stimulates water and ion secretion into the lumen of the large intestine (Racusen and Binder 1977; Waldman et al. 1977), and one study suggests that VIP effects were greater in the large than the small intestine (Wu et al. 1979). VIP increases chloride secretion and I_{sc} across monolayers of isolated human colonic epithlial cells (Dharmsathaphorn et al. 1984, 1985), indicating that, as in the small intestine, VIP receptors are present on the epithelium. These responses are closely correlated with ^{125}I-VIP binding and can be reduced by somatostatin or verapamil (Dharmsathaphorn et al. 1985) and the K^+-channel blocker quinidine (Cartwright et al. 1984); in isolated human colonic epithelial cells VIP effects can be reduced by adrenaline, via an a_2-receptor mechanism (Boige et al. 1984).

VIP causes an increase in intracellular cAMP levels (Simon et al. 1978; Dupont et al. 1980; Broyart et al. 1981), as in the small intestine. PHI, which has substantial structural homology with VIP, has similar actions on intestinal water and ion transport. This peptide interacts with VIP receptors, but with less affinity than VIP (Bataille et al. 1980; Laburthe et al. 1985). In rats (Ghiglione et al. 1982), pigs and humans (Anagnostides et al. 1983a,b; Moriarty et al. 1984) PHI causes either a decrease in absorption or an increase in secretion of water and ions.

7.3.2 Substance P

There have been comparatively few studies on the action of SP on mucosal transport in vivo. A possible secretory action of SP has been suggested in patients with carcinoid syndrome, in which tissue and serum levels of SP are elevated and diarrhea is a common symptom (Gamse et al. 1981); as with pancreatic cholera, however, other substances produced by the tumour (e.g. 5-HT) may contribute to these symptoms.

Small intestine. When injected intravenously into dogs, SP caused a profound secretion of water, sodium, chloride and potassium into the lumen of the proximal jejunum (McFadden et al. 1986), yet, in rats, SP stimulates absorption (Mitchenere et al. 1981). Under in vitro conditions, however, SP consistently causes an increase in I_{sc} in rats, guinea-pigs and rabbits (Walling et al. 1977; Kachur et al. 1982; Hubel et al. 1984; Keast et al. 1985a), which is mainly due to stimulation of chloride ion secretion, although neutral sodium chloride absorption is usually diminished.

These flux changes are not associated with any increase in epithelial cAMP levels (Walling et al. 1977; Laburthe et al. 1979), but are dependent on the presence of extracellular calcium (Kachur et al. 1982). Calcium is therefore a possible "second messenger" utilised in the secretory response to SP.

In guinea-pig and rabbit ileum the majority of the SP effect on I_{sc} was blocked by TTX (Hubel et al. 1984) and is therefore likely to be nerve-mediated. In our studies of guinea-pig ileal mucosa-submucosa, the major component of the I_{sc} response to SP was abolished by TTX and, of this nerve-mediated response, approximately 60% was hyoscine-sensitive, and hence probably involved stimulation of cholinergic submucous neurons (Keast et al. 1985a; Fig. 6); in these studies it was also shown that the SP analogue [D-arg^1, D-pro^2, D-trp^{7,9}, leu^{11}]-SP was able to reduce or abolish the TTX-resistant response to SP, but had no significant effect on the nerve-

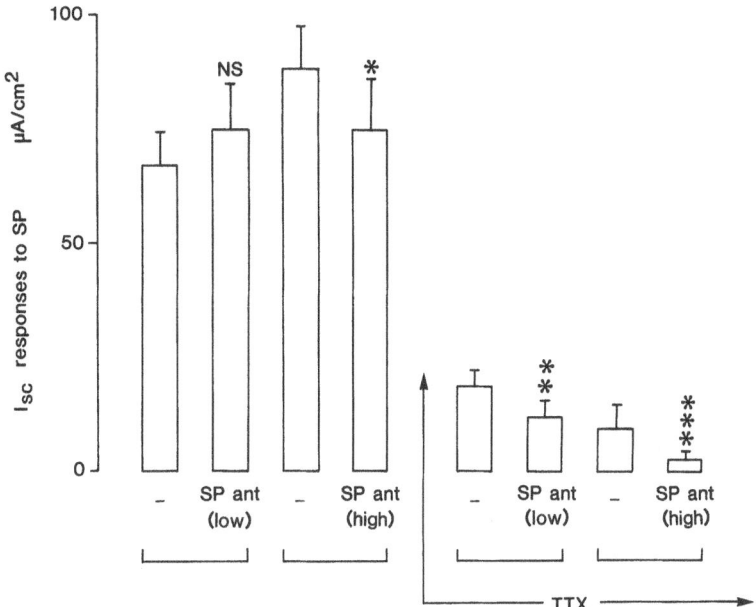

Fig. 6. Effect of the SP antagonist [D-arg[1], D-pro[2], D-trp[7, 9], leu[11]]-SP on I_{sc} induced by $10^{-7} M$ SP in the guinea-pig ileum mucosa-submucosa. Responses to SP in the absence and presence of TTX ($10^{-7} M$) are shown. Responses are expressed as $\mu A/cm^2$ exposed membrane. Antagonist concentrations are referred to as "low" ($6.7 \times 10^{-6} M$) or "high" ($3.4 \times 10^{-5} M$). The values represent mean ± SEM of at least 5 experiments (SP with low dose antagonist, $n = 8$); *$P < 0.05$; **$P < 0.025$; ***$P < 0.005$, compared to the appropriate control responses to SP (designated by bracket), paired two-tailed t-test. These results indicate that the majority of SP effects on I_{sc} are inhibited by TTX and are therefore nerve-dependent, and that the SP antagonist significantly diminishes the TTX-resistant component of the SP response, but has little or no effect on the major (nerve-mediated) component of the SP response. (Reprinted from Keast et al. (1985a) with minor modifications, with permission)

dependent component of the response (Fig. 6). This suggests that this analogue can be used to antagonize the action of SP on epithelial receptors, but is relatively ineffective on SP receptors on submucous neurons, in this species (Fig. 7).

Large intestine. SP had no effect on the I_{sc} of a human colonic tumour cell line, grown as a monolayer and perfused in a modified Ussing chamber (Dharmsathaphorn et al. 1984), which is consistent with SP having only a minor action directly on the epithelium, as suggested by some studies on the small intestine. As yet no other studies have examined the action of SP on the large intestine or the nature of the receptors involved.

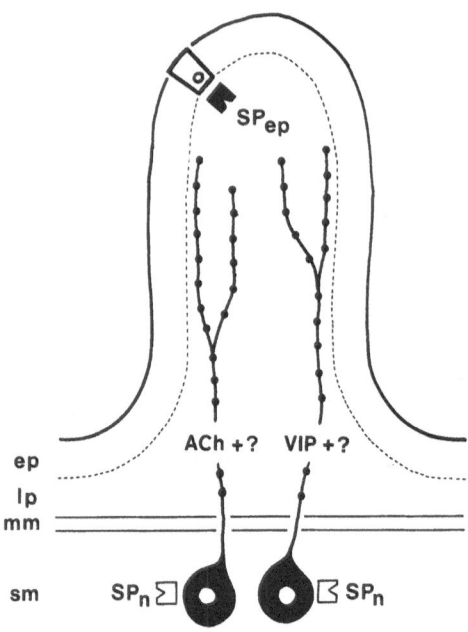

Fig. 7. Proposed location of SP receptors in the mucosa and submucosa of the guinea-pig small intestine. The layers of the intestine present in the Ussing chamber are represented (*ep*, epithelium; *lp*, lamina propria; *mm*, muscularis mucosae; *sm*, submucosa). Two types of SP receptors have been distinguished using the antagonist [D-arg^1, D-pro^2, D-trp$^{7,\ 9}$, leu^{11}]-SP. Those on submucous neurons are unaffected by the antagonist and have been designated SP_n, whereas those on epithelial cells (involved in the TTX-resistant responses to SP) are blocked by the antagonist and have been designated SP_{ep}. SP_n receptors are present on cholinergic (*ACh*) and non-cholinergic submucous neurons. In this species the non-cholinergic neurons contain *VIP*. The possibility of other active substances in these neurons is indicated. (Reprinted from Keast et al. (1985a) with minor modifications, with permission)

7.3.3 Somatostatin

Small intestine. When injected intravenously into rats, SOM stimulates fluid absorption (Dharmasathaphorn et al. 1980b; Mitchenere et al. 1981). An increase in coupled NaCl absorption and a decrease in bicarbonate secretion and I_{sc} are seen in vitro (Dharmsathaphorn et al. 1980a; Freedman et al. 1980; Guandalini et al. 1980; Kachur et al. 1980; Dobbins et al. 1981; Favus et al. 1981; Rosenthal et al. 1983). These in vitro effects are not altered by phentolamine, reserpine, atropine, carbachol or naloxone (Dharmsathaphorn et al. 1980a; Kachur et al. 1980), and do not appear to be accompanied by changes in epithelial cAMP levels (Laburthe et al. 1979; Dharmsathaphorn et al. 1980b). The effects of SOM on glucose and amino acid uptake are highly variable (Pott et al. 1979; Dharmsathaphorn et al. 1980a, b; Krejs et al. 1980; Märki 1981; Daumerie and Henquin 1982).

In experimental animals, the secretions caused by PGE_1, theophylline and cAMP are reduced or abolished by SOM (Dharmsathaphorn et al. 1980b; Dobbins et al. 1981). However, in normal, healthy humans there was no effect of SOM on basal absorption or VIP-induced secretion (Krejs et al. 1980). In contrast, in pancreatic cholera patients SOM enhanced absorption and caused a marked decline in the plasma VIP levels, for the period of SOM infusion (Davis et al. 1980; Krejs 1984); these antidiarrhoeal effects could also be produced by synthetic SOM analogues (Jaros et al. 1985; Santangelo et al. 1985). This may suggest a change in SOM receptor sensitivity during this disease or a dependence of SOM action on a background of activity or stimulant of secretion being present.

Until recently it has usually been assumed that the primary site of SOM action on mucosal transport was the epithelium. This has been investigated in the guinea-pig ileum, where SOM nerve terminals are found in the mucosa and submucosa, and where the major projection of submucous neurons is to the underlying mucosa (see above). It is therefore feasible that SOM could alter mucosal transport by reducing the activity of submucous secretomotor neurons. Using the isolated mucosa-submucosa, we have found that most of the I_{sc} decrease caused by SOM was abolished by TTX, suggesting that the major effect of SOM is dependent on nerve activity. Further experiments showed that exogenous SOM can inhibit the activity of both cholinergic and non-cholinergic submucous secretomotor neurons (Keast et al. 1986). As described for the experiments on the mechanism of action of NA on guinea-pig mucosa (see Sect. 7.2 and Keast et al. 1986), these studies do not exclude an effect of SOM on all epithelial cell transport processes. However, as the effect of TTX on the response to SOM was observed also in the presence of theophylline (a stimulant of chloride secretion), it is likely that SOM effects on chloride movement which alter I_{sc} are dependent on nerve activity.

Large intestine. SOM stimulates absorption of water, sodium and chloride, and causes a decrease in I_{sc} in the colon (Dobbins et al. 1981; Favus et al. 1981; Rosenthal et al. 1983). It also diminishes secretion evoked by 5-HT, theophylline, carbachol, PGE_1 and VIP (Carter et al. 1978; Dharmsathaphorn et al. 1980b; Guandalini et al. 1980; Dobbins et al. 1981). A reduction in the VIP-induced increase in I_{sc} was observed in a monolayer culture of human colonic tumour cells (Dharmsathaphorn et al. 1984), suggesting that SOM receptors exist on epithelial cells of this tissue. The possibility of a neural site of action of SOM has not been tested in the large intestine.

7.3.4 Neuropeptide Y and Related Peptides

In the rabbit ileum, NPY causes a decrease in I_{sc}, mainly due to an increase in chloride absorption; this effect was unchanged by TTX (Hubel and Renquist 1985). In this tissue, then, NPY probably acts only by a direct effect on the epithelium, rather than via nerves. Friel et al. (1986) have also shown that NPY causes a decrease in I_{sc} due to increased NaCl absorption and decreased Cl secretion, in isolated guinea-pig and rabbit ileum; PYY had similar effects as NPY, but was more potent, whereas rat PP had no effect on mucosal transport. A lack of effect of PP on mucosal transport has also been demonstrated in the small and large intestine of rats, rabbits and pigs (Gaginella et al. 1978b; Wu et al. 1979; Camilleri et al. 1981).

NPY has anti-secretory effects in the rat small intestine in vivo, where it also reduces VIP- and PGE_2-induced secretion; in this species PYY is less effective than NPY (Saria and Beubler 1985; McFadyen et al. 1986).

To my knowledge the actions of NPY and PYY on the mucosa of the large intestine have not been investigated.

7.3.5 Opiates and Opioid Peptides

The constipating effect of opiates is well known, and although some of this is due to an inhibition of intestinal motility, it is only quite recently that an additional, direct action on mucosal transport of secretomotor neurons has been acknowledged. This has been best illustrated by the many studies showing that opiates inhibit secretion or augment absorption, even in preparations of isolated mucosa-submucosa (see below).

Small intestine. Opiates and opioid peptides enhance absorption of water, sodium and chloride (represented by a decrease in I_{sc}, in vitro), as well as diminish secretion induced by VIP, bile salts, PGE_2, CT or ST (Valiulis and Long 1973; Coupar 1978, 1983; Beubler and Lembeck 1979, 1980; Dobbins et al. 1980, 1981; Kachur et al. 1980; Mailman 1980, 1984a; McKay et al. 1981; Sandhu et al. 1981; Ahrens and Zhu 1982b; Hughes et al. 1982, 1984; Kachur and Miller 1982; Turnberg et al. 1982; Barbezat and Reasbeck 1983; Vinayek et al. 1983, 1985; Warhurst et al. 1983; Brown and Miller 1984; Farack and Loeschke 1984; Fogel and Kaplan 1984); they are ineffective against mannitol-induced secretion (Beubler and Lembeck 1979). These anti-secretory effects are probably not associated with any change in epithelial cAMP levels (Laburthe et al. 1979; Hardcastle et al. 1981a). Attempts to define opiate receptors on intestinal epithelial cells have had mixed success (Gaginella et al. 1983; Binder et al. 1984; López-Ruiz et al. 1985). Where various types of opiates have been used in the same tissue, δ-agonists appear to be the most effective. Dynorphin $(1-13)$, dermorphin and β-endorphin cause much smaller decreases in I_{sc} (Kachur and Miller 1982).

In rabbits the effects of enkephalins on transport and I_{sc} are completely blocked by TTX but are unaffected by atropine, phentolamine, haloperidol, yohimbine or sympathectomy (Dobbins et al. 1980; Turnberg et al. 1982;

Binder et al. 1984). This suggests that, in this species, enkephalins do not have a significant effect on epithelial cells, but instead act on intrinsic non-cholinergic secretomotor neurons. This is likely to be via inhibition of nerves which cause secretion, rather than stimulation of nerves which cause absorption, as no intrinsic nerve population has yet been found which, when activated, causes net absorption (see above). This neural site of action is consistent with electrophysiological studies in the guinea-pig ileum, which showed that opiates hyperpolarize submucous neurons, by an action on δ-receptors (Mihara and North 1986).

Administration of naloxone or naltrexone in vivo (but not in vitro) to morphine-dependent rats elicits a "withdrawal" phenomenon, a component of which is profound secretion of water and electrolytes (Warhurst et al. 1984); the effect of naltrexone can be partly reduced by atropine or hexamethonium (Beubler et al. 1984; Chang et al. 1984). Naloxone alone usually has no effect on absorption in non-dependent animals, although Fogel and Kaplan (1984) reported that it decreased absorption in the rat small intestine, an effect which was inhibited by atropine.

The nerve-dependent effects of exogenous opioid peptides in rabbits are consistent with the distribution of some endogenous opioid peptides in this species, in which nerve fibres containing ENK are very rare or absent in the mucosa, but ENK nerve terminals near submucous neurons (which probably project to the mucosa) are found. This distribution of ENK is similar to that observed in other species (see Sect. 4.3.5), indicating that the epithelial cells are unlikely to be exposed to significant concentrations of ENK. However, the distribution of other opioid peptides (e.g. dynorphin) in the mucosa has not been studied in rabbits, and it is possible that these may act directly on epithelial cells.

Large intestine. Opiates and opioid peptides have similar effects here as in the small intestine, stimulating basal absorption and reducing the secretory effects of bile salts, VIP, PGE_1 and CT (Gordon et al. 1978; Beubler and Lembeck 1979; Farack et al. 1981; Warhurst et al. 1983; Farack and Loeschke 1984), but not affecting that caused by mannitol (Beubler and Lembeck 1979). The effect on absorption may not be directly on the epithelium, as met-ENK had no effect on the I_{sc} of a human colonic tumour epithelial cell culture (Dharmsathaphorn et al. 1984). The effect of TTX on opiate-induced absorption has not been studied in the large intestine.

7.4 Summary

There are many substances found in endocrine or neural tissues of the gastrointestinal tract which have been found to affect mucosal transport and the majority of these have a net secretory action (listed by Tapper 1983). Some generalizations can be made about the substances which have been discussed so far. Stimulation of the vagus or mucosal nerve fibres containing acetylcholine, SP or VIP would be expected to cause secretion. Inhibition of secretion could be initiated at the mucosal level by release of SOM or NPY; however, the action of ENK or NA is likely to be primarily by release from nerve terminals in submucous ganglia. SP and SOM possibly have two physiologically relevant sites of action, the epithelium and submucous neurons (Table 1).

8 Stimulation of Mucosal Nerve Fibres In Vivo

In 1859 Claude Bernard reported that surgical removal of the mesenteric ganglia in dogs led to an enhanced intestinal secretion and diarrhoea. In subsequent years these experiments were repeated, with similar results (see reviews by Florey et al. 1941; Babkin 1950). The secretion observed was called "paralytic secretion" and was abolished by atropine (Hanau 1886; Molnár 1909); however, an atropine-resistant component of this secretion was observed by Brunton and Pye-Smith (1876). In other early experiments secretion was elicited by pilocarpine (Reid 1892), vagal stimulation (Savitch and Sochestvensky 1917) or feeding (Molnár 1909) and absorption was elicited by atropine alone (Molnár 1909; Rabinovitch 1927). More recent studies with atropine have also demonstrated its absorptive action in vivo (Blickenstaff and Lewis 1952; Tidball and Tidball 1958; Hubel 1976; Ahrens and Zhu 1982a) and have shown that this action is not accounted for by motility or cardiovascular changes (Morris and Turnberg 1980; Mailman 1984b). The conclusion from these studies is that there are nerve pathways from extrinsic ganglia to the intestine which can influence mucosal function, the vagal cholinergic (and possibly non-cholinergic) input causing net water and ion secretion. The nerve pathways causing secretion appeared to be continuously active and under tonic inhibition from sympathetic nerves. Brunton and Pye-Smith (1876) also put forward the idea of intrinsic secretomotor neurons, although experimental evidence at that stage was lacking.

A more extensive study of nerve-mediated secretion was carried out by Wright et al. (1940). Decerebrated or decapitated cats were used, in order to avoid the use of anaesthetics, which were found to depress both basal and evoked secretion. Under these conditions vagal stimulation caused a profound increase in the volume of secretion from the duodenum, but this consisted mostly of mucus and was thought to have come from Brunner's glands. Neither vagal stimulation nor vagotomy had an effect on the volume of secretion produced by the lower small intestine; moreover, stimulation of the pelvic nerves, which enhanced colonic secretion (Wright et al. 1938), had no effect on secretion by the small intestine. However, after either administration of eserine or cutting all of the preganglionic sympathetic fibres, or vagal stimulation after section of the greater splanchnic nerve, secretion throughout the small intestine was observed. The secretory responses were atropine-sensitive after eserine or sympathetic nerve section, remained after vagotomy and could be attenuated by stimulating the distal end of the severed splanchnic nerves. The exact nature of the secretion is not known; it undoubtedly contained mucus, but may also have contained watery secretion from the crypts. Taken together, these experiments suggest that, throughout the small intestine, there are tonically active, intrinsic cholinergic neurons which enhance secretion and that vagal activity excites these intrinsic neurons; the effect of the vagus is usually masked by the activity of inhibitory sympathetic nerve fibres, an activity which is particularly marked in the lower small intestine. This is consistent with the results from more recent studies in which vagotomy had no effect on basal absorption and secretion (Tidball and Tidball 1958; Bunch and Shields 1973).

Similar experiments have been carried out on the large intestine of cats, as summarized by Wright et al. (1938) and Florey et al. (1941). Stimulation of the pelvic nerves enhanced secretion, an effect which was blocked by atropine and potentiated by eserine; pilocarpine mimicked the effect of pelvic nerve stimulation. Section of the sympathetic nerves did not cause a "paralytic secretion" (as seen in the small intestine), but stimulation of these nerves reduced the secretory effect of pelvic nerve stimulation. The parasympathetic input appears, then, to have a greater influence (and/or the sympathetic input a smaller influence) in the large than in the small intestine.

The possible role of the nervous system in controlling intestinal absorption and secretion in vivo was not investigated further until the past few years, when a very thorough series of experiments to demonstrate actions of enteric secretomotor neurons in vivo was carried out by Sjövall, Lundgren and co-workers. Fluid absorption from cannulated loops of small intestine of cats and rats was monitored and, because of the potentially complex interactions of other intestinal processes with mucosal transport, vascular and motility changes were either carefully monitored or minimized. This work showed that, in animals with severed splanchnic nerves, vagal stimulation caused an atropine-resistant fluid secretion, and that atropine itself increased fluid absorption (Sjövall et al. 1983a). These results suggest that there are both non-cholinergic and cholinergic secretomotor neurons, and that these cholinergic neurons are continuously active. However, atropine generally has no significant effect on basal I_{sc} in vitro, which suggests that the intrinsic cholinergic secretomotor neurons are not active under these conditions.

Intrinsic cholinergic secretomotor nerves have also been demonstrated in surgically isolated or denervated segments of intestine. Nassett et al. (1935)

showed that the secretory response to mechanical stimulation of the mucosa can be evoked in transplanted or isolated intestinal segments, in which all extrinsic neural inputs had been severed. Knaffl-Lenz and Nagaki (1925) showed that placing a hypertonic solution into the lumen of an isolated intestinal loop caused secretion, and that this effect was blocked by atropine and augmented by pilocarpine. More recently, Caren et al. (1974) showed that the secretion evoked in extrinsically denervated Thiry-Vella loops of dog intestine by luminal distension or tactile stimulation could be blocked with atropine or hexamethonium.

Recent studies support the earlier conclusions that sympathetic nerve stimulation enhances absorption. An increase in absorption can be elicited in vivo in the small intestine of dogs by "electrical pacing" (i.e. directly passing currents across the whole thickness of gut wall). This absorption was not dependent on motility changes and was blocked by phentolamine, but not propranolol (Collin et al. 1979; Björck et al. 1984). This technique appears to stimulate the sympathetic nerve fibres in the intestine, leading to an increase in absorption. Other studies carried out on atropinized cats also demonstrated the absorptive effects of splanchnic nerve stimulation, which were blocked by phentolamine and mimicked by noradrenaline (Brunsson et al. 1979). These apparent noradrenergic effects were attributed primarily to the inhibition of secretion in the crypts (Sjövall et al. 1983 b). It has also been shown that splanchnic nerve stimulation activated two distinct nerve pathways, one which caused absorption (which was hexamethonium-resistant) and another which caused vasoconstriction (which was blocked by hexamethonium; Sjövall 1984a). This idea of two distinct sympathetic nerve pathways had already been proposed much earlier by Brunton and Pye-Smith (1876), who noticed that, after cutting the mesenteric nerves, the hemorrhagic and secretory effects usually had different time courses. Recent immunohistochemical studies (Costa and Furness 1984; Macrae et al. 1986) have shown that the noradrenergic neurons that supply intestinal arterioles and submucous ganglia can also be distinguished by their chemical coding.

The tonic absorptive or anti-secretory effects of sympathetic nerves have also been demonstrated by Chang et al. (1985), who showed that rats with degenerated noradrenergic nerve terminals (after either streptozocin or 6-OHDA pretreatment) have diminished ileal water and sodium absorption; this decrease was not apparent in the jejunum, which may suggest different secretomotor innervation at that site. However, adrenoreceptor antagonists alone generally have no effect on basal or stimulated ion movement in vitro (Chang et al. 1982), which implies that noradrenaline is not being released continuously to alter mucosal transport when noradrenergic terminals are separated from their cell bodies.

9 Microcircuitry of Secretomotor Pathways

It has now been established that stimulation of submucous neurons and their mucosal processes both in vivo and in vitro causes changes in mucosal water and ion movement. As these submucous neurons are the source of the majority of mucosal nerve fibres, they are likely to be a primary site of integration for secretomotor reflexes. Important insights into the microcircuitry of submucous ganglia have been obtained from intracellular microelectrode studies of these neurons in guinea-pig small intestine. Submucous neurons can be classified on the basis of the types of synaptic input they receive (Hirst and McKirdy 1975; Surprenant 1984a, b; North and Surprenant 1985; Bornstein et al. 1986, 1987). A small number of cells appear to have no synaptic input, but in all of the remaining cells fast excitatory synaptic potentials (ESPs) can be evoked. Of these cells, some also exhibit slow ESPs. Fast ESPs are thought to be mediated by acetylcholine, via nicotinic receptors, whereas the transmitter involved in the slow ESP has not been identified. Some of the fast and slow ESPs are produced by nerve terminals which arise from myenteric ganglia (Bornstein et al. 1987).

Cholinergic secretomotor neurons and interneurons have been defined physiologically (see Sect. 8). In the guinea-pig small intestine, three populations of submucous cholinergic neurons can be identified histochemically (Furness et al. 1984, 1985). In this species the ChAT/CCK/CGRP/NPY/SOM and ChAT/SP neurons may be secretomotor neurons, as their processes are found in the mucosa, but they are unlikely to be interneurons, as there are very few NPY nerve terminals in submucous ganglia (Furness et al. 1983b) and no SP nerve terminals remain in submucous ganglia after myectomy and extrinsic denervation (Costa et al. 1981). Thus, it is likely that the cholinergic neurons which contain no other peptide are the source of ESPs which remain after myectomy (Bornstein et al. 1987) and these neurons might be interneurons in secretomotor reflexes. The population of myenteric neurons which provide excitatory inputs to submucous neurons has not been identified structurally.

Inhibitory synaptic potentials (ISPs) are observed in those submucous neurons which exhibit slow ESPs (Bornstein et al. 1986). It has been suggested that NA is the transmitter responsible for the ISPs as a_2-antagonists or 6-OHDA added in vitro abolish the ISPs, and NA mimics the ISP only in those neurons in which ISPs can be evoked (North and Surprenant 1985). Recent studies have shown that, in the guinea-pig, noradrenergic nerve terminals preferentially innervate non-cholinergic submucous neurons (i.e. the VIP/DYN neurons) and that ISPs can only be evoked in these neurons (Bornstein et al. 1986). These studies are consistent with mucosal transport studies, in which NA reduced the activity of non-cholinergic submucous secretomotor neurons, as stimulated by EFS or low concentrations of 5-HT (Keast et al.

1986); the inhibition by NA of the cholinergic secretory responses may indicate that NA can reduce the activity of these neurons by acting on receptors on initial or terminal regions of their axons (Keast et al. 1986).

A population of ISPs persists after extrinsic denervation (i.e. when noradrenergic nerve terminals are no longer present; Hirst and McKirdy 1975; J. C. Bornstein, J. B. Furness and M. Costa, unpublished observations); however, the transmitter involved has not been determined. Likely candidates are somatostatin or an opioid peptide, as both hyperpolarize submucous neurons (Mihara and North 1986), both are present in intrinsic neurons that supply submucous ganglia (Costa et al. 1980; Furness et al. 1983c) and both inhibit water and ion secretion by a TTX-sensitive mechanism (see above).

10 Functions of Secretomotor Neurons

10.1 Types of Secretomotor Reflexes

As textbook descriptions attest, until quite recently the regulation of transmucosal ion movement had been attributed almost solely to the properties of the epithelium, conditions in the underlying interstitium, mucosal blood flow and hormonal influences. Nevertheless, neural secretomotor reflexes have been demonstrated in many recent experiments, particularly in cats. For example, a secretomotor reflex can be elicited by intraluminal glucose (Sjövall et al. 1983c; Sjövall 1984b). In these experiments NA or stimulation of the sympathetic nerves caused absorption only when glucose was present in the luminal perfusate. As the NA effect was TTX-sensitive and TTX itself caused a marked net absorptive change when glucose was present in the lumen, it was proposed that both sympathetic nerve activity and NA cause net absorption by inhibiting a glucose-activated secretory reflex. A similar secretory reflex may be activated by alanine or polyethylene glycol (Sjövall et al. 1984a). A possible physiological role of this glucose-activated secretory reflex has been suggested. Glucose, water and electrolytes are absorbed across the villus epithelium and, by activation of this reflex, water and electrolytes are secreted across the crypt epithelium; thus, the consequential uptake of water and electrolytes with glucose is reduced.

Bacterial toxins, in particular cholera toxin (CT), also elicit secretomotor reflexes and have been studied in considerable detail, again mainly in cats. As CT appears to bind to and penetrate only the villus cells (Hansson et al. 1984), and as CT causes secretion primarily by stimulating active secretion, an indirect mechanism for stimulation of crypt cells is predicted. The studies of Lönnroth and Jennische (1982) showed that a variety of anaesthetics and sedatives inhibited CT-induced secretion. Other studies have shown that the CT secretory response was considerably reduced by TTX, lidocaine and hexamethonium (Cassuto et al. 1981 b, 1983); in addition, CT caused a TTX-sensitive release of VIP from the secreting intestinal segment (Cassuto et al. 1981 a) and concurrent degranulation of 5-HT from enterochromaffin cells (Nilsson et al. 1983). The secretion was mimicked by intravenous VIP or 5-HT, or intraluminal 5-HT, with the VIP response unaffected by either TTX or hexamethonium (Cassuto et al. 1983), while the responses to 5-HT were reduced by either agent (Cassuto et al. 1982a). It has therefore been suggested that at least some of the CT-evoked secretion is mediated by a neuronal reflex; the proposed theory is that the toxin may bind to enterochromaffin cells, causing release of 5-HT onto nearby mucosal nerve endings, initiating a secretory reflex. The nervous mechanism appears to involve a nicotinic synapse, and the final effector neuron may release VIP. Extrinsic nerves may be involved in the modulation of this response as the secretion can be reduced by sympathetic nerve stimulation (Cassuto et al. 1982b). A similar neuronal reflex has been proposed as a mechanism of action of *E. coli* enterotoxin (Eklund et al. 1985). In pigs, the secretion induced by this toxin may involve cholinergic neurons, as the response is reduced by atropine (Ahrens and Zhu 1982b).

As yet it has not been possible to demonstrate in vitro that CT-evoked secretion occurs by a neural mechanism. In muscle-stripped preparations of isolated guinea-pig or rabbit ileum, CT secretion is unaltered by TTX (Cooke and Carey 1984; Moriarty et al. 1985) or, in the guinea-pig, by the 5-HT antagonist cisapride (Cooke and Carey 1984); this may represent either a species difference in the mechanism of CT action or a disruption in the reflex pathway during tissue removal and dissection (e.g. the reflex may require pathways that pass through the myenteric plexus, which was removed in these experiments).

Secretomotor reflexes can also be elicited by mechanical stimulation of the mucosa (Nassett et al. 1935; Caren et al. 1974) or by intraluminal application of hypertonic solutions (Knaffl-Lenz and Nagaki 1925). More recently luminal perfusion with bile salts (Karlström et al. 1983), dibutyryl-cAMP or theophylline (Eklund et al. 1984), or serosal application of dilute hydrochloric acid (Sjöqvist et al. 1982), bile acids or ethanol (Brunsson et al. 1985) have been shown to cause secretion in the jejunum of cats and rats. In these later studies, the secretion was reduced by close intra-arterial injection of TTX, serosal application of lidocaine or intravenous hexamethonium. Secretory reflexes were elicited in both control and peri-arterially denervated segments and were unaffected by atropine. Taken together, these studies suggest that the chemically induced secretion was mediated by a reflex involving a nicotinic synapse and a non-cholinergic final neuron.

The above studies on bile salt-induced secretion differ slightly from those of Kvietys et al. (1979) in rabbits, in which the secretion was atropine-sensitive; this difference may reflect interspecies variation in the intramural nerve pathways. Taub et al. (1977) and Coyne et al. (1977) observed that bile acid-induced secretion was blocked by propranolol, which, in high concentrations is thought to have an anaesthetic effect; this may indicate that the secretion was dependent on neural activity.

The mechanism by which sensory nerve endings are stimulated in these secretory reflexes and the transmitters utilised by sensory nerve fibres are not known. The studies of Brunsson et al. (1985; see above) have suggested that the secretion induced by serosal irritation (as may be evoked in peritonitis) may stimulate sensory nerve endings via the release of the PGs, bradykinin or histamine. It is possible that 5-HT, released by enterochromaffin cells during CT exposure, is the common stimulant of the sensory nerve endings in other secretomotor reflexes. The cell bodies of sensory neurons involved in intrinsic reflexes must be in the myenteric or submucous plexuses; however, there might also be sensory endings of extrinsic sensory neurons in the mucosa.

Under some circumstances neural mechanisms for the stimulation of absorption rather than secretion appear to be activated. Bilateral carotid occlusion, which activates baroreceptors and stimulates peripheral sympathetic nerves, also increases intestinal fluid uptake (Sjövall et al. 1982). Cutting the mesenteric nerves, as well as causing a decrease in the resting absorption level, inhibited the effect of occlusion. The sympathetic nerves may therefore contribute to a reflex compensatory mechanism to regulate extracellular fluid volume (e.g. in cases of hypovolemic shock or hemorrhage; Sjövall et al. 1982; Redfors and Sjövall 1984). It has been suggested that this mechanism is mediated by elevated angiotensin II levels, which then enhance the release from noradrenergic nerves (Levens 1985).

10.2 General Organization of Secretomotor Reflexes

A general arrangement of secretomotor reflexes can be postulated, as shown in Fig. 8. This schema is based on histochemical and physiological information derived mainly from guinea-pig tissues in vitro, whereas in vivo evidence for these reflexes comes mainly from dogs and cats, where the microcircuitry may differ. There are both cholinergic and non-cholinergic secretomotor neurons. There are also sensory nerve fibres in the mucosa, responsive to stimuli such as glucose and bacterial toxins; some of the sensory fibres arise from intrinsic (submucous or myenteric) neurons and some may also come from extrinsic ganglia. There are probably interneurons in the intrinsic reflexes; these could be in either submucous or myenteric plexuses. Secretomotor reflexes may, when appropriate, interact with reflexes controlling muscular activity.

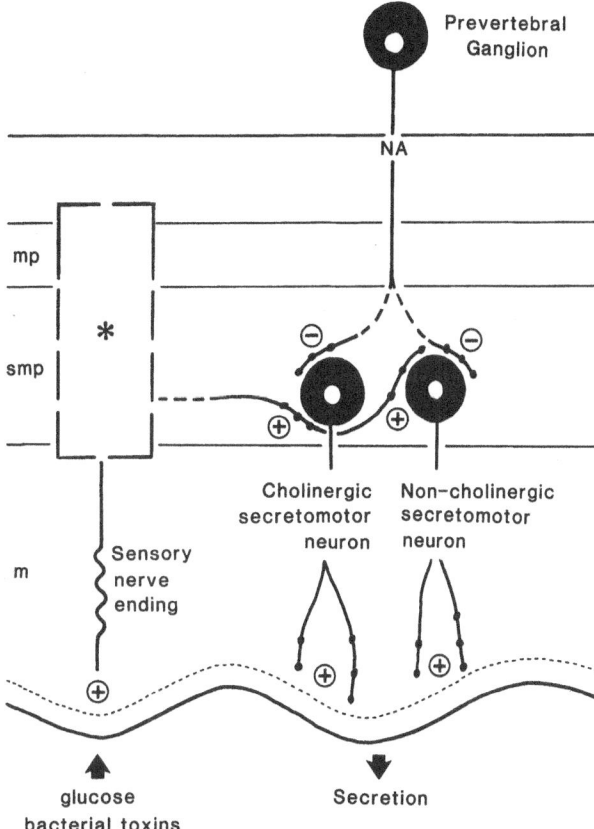

Fig. 8. Components of secretomotor reflexes in the small intestine. Stimuli such as intra-luminal glucose or bacterial toxins act on sensory nerve endings in the mucosa. The locations of these sensory fibres, their cell bodies and those of subsequent interneurons in the reflex have not been defined (indicated by *broken lines* and *asterisks*). Activation of this pathway causes stimulation of cholinergic and non-cholinergic secretomotor neurons in the submucosa, leading to an enhanced secretion of water and electrolytes. Nerve terminals from sympathetic noradrenergic neurons can reduce the activity of the final secretomotor neurons. In some species (such as guinea-pigs), only the non-cholinergic neurons are innervated by noradrenergic fibres, whereas in some other species (e.g. cats) there is functional evidence that both cholinergic and non-cholinergic neurons are inhibited by sympathetic innervation

A question yet to be investigated is the physiological relevance of two (or more) active substances coexisting within the one submucous neuron. In the submucosa some neurons contain substances which have similar actions on mucosal transport (e.g. NA and SOM cause absorption, ACh and SP cause secretion), whereas other substances that are found together have opposing actions (e.g. ACh causes secretion, but SOM and NPY cause absorption). Preliminary studies in guinea-pig small intestine suggest that the effects of NA and SOM on mucosal transport, which are in the same neurons, are no more than additive (Keast et al. 1986); however, other combinations of substances or other types of interactions (e.g. effects on transmitter release) have not been investigated.

Immunohistochemical studies have suggested that there are a number of possible target tissues for mucosal nerve fibres, and interactions between blood flow, smooth muscle contraction (in external muscle and muscularis mucosae), hormonal secretion (from endocrine cells) and water and ion fluxes are expected. These interactions have not yet been defined. A considerable amount is now known about the neural circuitry and the secretory effects of some substances contained in mucosal nerve fibres, particularly in guinea-pig small intestine. It is now important to examine in more detail in this species the mechanisms of action of substances (e.g. glucose, bacterial toxins) which elicit secretomotor reflexes on other species and to determine the microscopic anatomy of the circuitry in species such as cats and dogs, where physiological studies have been made.

10.3 Physiological Roles of Secretomotor Reflexes

There are several physiological situations where neural mechanisms may modify mucosal water and electrolyte transport. For example, in the presence of nutrients (e.g. sugars, amino acids) stimulation of fluid secretion by the crypts would allow for selective uptake of the nutrients, if the secreted fluid and electrolytes are reabsorbed with nutrients in the villi. There is experimental evidence for a glucose-elicited secretory reflex in cats (Sjövall et al. 1983c, 1984a), but the physiological importance of this has not yet been demonstrated. As the absorption of water, electrolytes and nutrients varies along the gastrointestinal tract, there may be corresponding changes in secretomotor reflexes. Moreover, there may be reflexes which coordinate the events between different intestinal sites.

Some secretomotor reflexes can be thought of as "local" reflexes, in that they directly pertain to intestinal functions. Tonic activity of secretomotor neurons in the mucosa and submucosa (as demonstrated by the effects of TTX) provides a mechanism for changes in the luminal contents to rapidly increase or decrease secretomotor neuron activity and therefore permit fine control over absorptive and digestive processes.

From the previous discussion (see Sects. 8 and 10.1) it is clear that mucosal transport can also be modified by extrinsic nerves. It is possible that much broader controls of mucosal transport are provided by these pathways to coordinate intestinal absorption and secretion with other body functions, notably whole body water and electrolyte balance. There is a considerable amount of evidence for interactions of the renin-angiotensin system with intestinal transport and the enteric nervous system (as summarized by Levens 1985) and experiments by Sjövall et al. (1982, 1984b) indicate that activation of cardiovascular mechano- or baroreceptors can also alter intestinal fluid transport.

The functional innervation of the mucosa should therefore be considered not only in terms of regulating the local intestinal environment, but also as a potentially vital component of the body's regulatory system for maintaining water and electrolyte homeostasis.

Acknowledgements. I wish to thank Professors John Furness and Marcello Costa for their expert advice during the experiments I carried out on mucosal innervation. These experiments were supported by a Commonwealth Postgraduate Research Award and the National Health and Medical Research Council of Australia. I am also very grateful to Professor Furness for his thorough and constructive criticism of this manuscript.

References

Ahrens FA, Zhu B (1982a) Effects of indomethacin, acetazolamide, ethacrynate sodium, and atropine on intestinal secretion mediated by *Escherichia coli* heat-stable enterotoxin in pig jejunum. Can J Physiol Pharmacol 60:1281–1286

Ahrens FA, Zhu B (1982b) Effects of epinephrine, clonidine, L-phenyl-ephrine and morphine on intestinal secretion mediated by *Escherichia coli* heat-stable enterotoxin in pig jejunum. Can J Physiol Pharmacol 60:1680–1685

Anagnostides AA, Christofides ND, Tatemoto K, Bloom SR, Chadwick VS (1983a) Secretory effects of peptide histidine isoleucine (PHI) in human jejunum. Scand J Gastroenterol 18 [Suppl 87]:112–113

Anagnostides AA, Manola K, Christofides ND, Yiangou Y, Welbourn RB, Bloom SR, Chadwick VS (1983b) Peptide histidine isoleucine (PHI). A secretagogue in porcine intestine. Dig Dis Sci 28:893–896

Andres H, Bock R, Bridges RJ, Rummel W, Schreiner J (1985) Submucosal plexus and electrolyte transport across rat colonic mucosa. J Physiol (Lond) 364:301–312

Aulesbrook KA (1965a) Intestinal absorption of glucose and sodium: effects of epinephrine and norepinephrine. Biochem Biophys Res Commun 18:165–169

Aulesbrook KA (1965b) Intestinal transport of glucose and sodium: stimulation by reserpine and the humoral mechanisms involved. Proc Soc Exp Biol Med 119:387–389

Babkin BP (1950) Secretory mechanism of the digestive glands, 2nd edn. Hoeber, New York, pp 614–632

Baldissera FGA, Holst JJ, Jensen SL, Krarup T (1985) Distribution and molecular forms of peptides containing somatostatin immunodeterminants in extracts from the entire gastrointestinal tract of man and pig. Biochim Biophys Acta 838:132–143

Barbezat GO, Grossman MI (1971) Intestinal secretion: stimulation by peptides. Science 174:422–424

Barbezat GD, Reasbeck PG (1983) Effects of bombesin, calcitonin and enkephalin on canine jejunal water and electrolyte transport. Dig Dis Sci 28:273–277

Baskin DG, Ensinck JW (1984) Somatostatin in epithelial cells of intestinal mucosa is present primarily as somatostatin 28. Peptides 5:615–621

Bataille D, Gespach C, Laburthe M, Amiranoff B, Tatemoto K, Vaculin N, Mutt V, Rosselin G (1980) Porcine peptide having N-terminal histidine and C-terminal isoleucine (PHI). Vasoactive intestinal peptide (VIP) and secretin-like effect in different tissues from the rat. FEBS Lett 114:240–242

Baumgarten HG (1967) Über die Verteilung von Catecholaminen im Darm des Menschen. Z Zellforsch 83:133–146

Berkley HJ (1893) The nerves and nerve endings of the mucous layer of the ileum, as shown by the rapid Golgi method. Anat Anz 8:12–19

Bernard C (1859) Leçons sur les liquides de l'organisme. Ballière, Paris

Bernard C (1864) Du rôle des actions réflexes paralysantes dans les phénomènes des sécrétions. J Anat (Paris) 1:507–513

Beubler E (1980) Influence of vasoactive intestinal polypeptide on net water flux and cyclic adenosine 3′,5′-monophosphate formation in the rat jejunum. Naunyn-Schmiedeberg's Arch Pharmacol 313:243–247

Beubler E (1981) VIP and PGE$_1$ activate adenylate cyclase in rat intestinal epithelial cell membranes via different mechanisms. Eur J Pharmacol 74:67–72

Beubler E, Lembeck F (1979) Inhibition of stimulated fluid secretion in the rat small and large intestine by opiate agonists. Naunyn-Schmiedeberg's Arch Pharmacol 306:113–118

Beubler E, Lembeck F (1980) Inhibition by morphine of prostaglandin-E$_1$-stimulated secretion and cyclic adenosine 3′,5′-monophosphate formation in the rat jejunum in vivo. Br J Pharmacol 68:513–518

Beubler E, Bukhave K, Rask-Madsen J (1984) Colonic secretion mediated by prostaglandin-E$_2$ and 5-hydroxytryptamine may contribute to diarrhea due to morphine withdrawal in the rat. Gastroenterology 87:1042–1048

Biber B, Lundgren O, Svanvik J (1971) Studies on the intestinal vasodilatation observed after mechanical stimulation of the mucosa of the gut. Acta Physiol Scand 82:177–190

Billroth H (1858) Einige Beobachtungen über das ausgedehnte Vorkommen von Nervenanastomosen im Tractus intestinalis. Arch Anat Physiol (Leipzig) 2:148–158

Binder HJ, Laurensen JP, Dobbins JW (1984) Role of opiate receptors in regulation of enkephalin stimulation of active sodium and chloride absorption. Am J Physiol 247:G432–G436

Björck S, Phillips SF, Kelly KA (1984) Mechanisms of enhanced intestinal absorption with electrical pacing. Gastroenterology 86:1029

Blickenstaff DD, Lewis LJ (1952) Effect of atropine on intestinal absorption of water and chloride. Am J Physiol 170:17–23

Bloom SR, Polak JM, Pearse AGE (1973) Vasoactive intestinal polypeptide and watery-diarrhea syndrome. Lancet II:14–16

Bloom SR, Delamarter J, Kawashima E, Christofides ND, Buell G, Polak JM (1983) Diarrhoea in VIPoma patients associated with cosecretion of a second active peptide (peptide histidine isoleucine) explained by a single coding gene. Lancet II:1163–1165

Boige N, Munck A, Laburthe M (1984) Adrenergic versus VIPergic control of cyclic AMP in human colonic crypts. Peptides 5:379–383

Bolton J, Field M (1977) Calcium ionophore-stimulated ion secretion in rabbit ileal mucosa: relation to actions of cyclic AMP and carbamylcholine. J Membr Biol 35:159–173

Bornstein JC, Costa M, Furness JB (1986) Synaptic inputs to immunohistochemically identified neurones in the submucous plexus of the guinea-pig small intestine. J Physiol (Lond) 381:465–482

Bornstein JC, Furness JB, Costa M (1987) Sources of synaptic inputs to neurochemically indentified submucous neurones of guinea-pig small intestine. J Auton Nerv Syst 18:83–91

Breiter W, Frey H (1862) Zur Kenntnis der Ganglien der Darmwand des Menschen. Z Wiss Zool 11:125–134

Brodin E, Sjölund K, Håkanson R, Sundler F (1983) Substance P-containing nerve fibers are numerous in human but not in feline intestinal mucosa. Gastroenterology 85:557–564

Brown DR, Miller RJ (1984) Adrenergic mediation of the intestinal antisecretory action of opiates administered into the central nervous system. J Pharmacol Exp Ther 231:114–119

Browning JG, Hardcastle J, Hardcastle PT, Sanford PA (1977) The role of acetylcholine in the regulation of ion transport by rat colon mucosa. J Physiol (Lond) 272:737–754

Browning JG, Hardcastle J, Hardcastle PT, Redfern JS (1978) Localization of the effect of acetylcholine in regulating intestinal ion transport. J Physiol (Lond) 281:15–27

Broyart JP, Dupont C, Laburthe M, Rosselin G (1981) Characterization of vasoactive intestinal peptide receptors in human colonic epithelial cells. J Clin Endocrinol Metab 52:715–721

Brunemeier EH, Carlson AJ (1914) Contributions to the physiology of the stomach. XIX. Reflexes from the intestinal mucosa to the stomach. Am J Physiol 36:191–195

Brunnson I, Eklund S, Jodal M, Lundgren O, Sjövall H (1979) The effect of vasodilatation and sympathetic nerve activation on net water absorption in the cat's small intestine. Acta Physiol Scand 106:61–68

Brunsson I, Sjöqvist A, Jodal M, Lundgren O (1985) Mechanisms underlying the intestinal fluid secretion evoked by nociceptive serosal stimulation of the rat. Naunyn-Schmiedeberg's Arch Pharmacol 328:439–445

Brunton TL, Pye-Smith PH (1876) The conditions of intestinal secretion and movement. Br Assn Adv Sci, John Murray, London, pp 308–314

Bryant MG, Polak JM, Modlin I, Bloom SR, Albuquerque RH, Pearse AGE (1976) Possible dual role for vasoactive intestinal peptide as gastrointestinal hormone and neurotransmitter substance. Lancet I:991–993

Bülbring E, Lin RCY, Schofield G (1958) An investigation of the peristaltic reflex in relation to anatomical observations. Q J Exp Physiol 43:26–37

Bunce KT, Spraggs CF (1983) a_2-Adrenoceptors mediate the antisecretory effects of a-adrenoceptor agonists in rat jejunum. Scand J Gastroenterol 18 [Suppl 87]:105

Bunch GA, Shields R (1973) The effects of vagotomy on intestinal handling of water and electrolytes. Gut 14:116–119

Camilleri M, Cooper BT, Adrian TE, Bloom SR, Chadwick VS (1981) Effects of vasoactive intestinal peptide and pancreatic polypeptide in rabbit intestine. Gut 22:14–18

Caren JF, Meyer JH, Grossman MI (1974) Canine intestinal secretion during and after rapid distension of the small bowel. Am J Physiol 227:183–188

Carey HV, Cooke HJ, Zafirova M (1985) Mucosal responses evoked by stimulation of ganglion cell somas in the submucosal plexus of the guinea-pig ileum. J Physiol (Lond) 364:69–80

Carter RF, Bitar KN, Zfass AM, Makhlouf GM (1978) Inhibition of VIP-stimulated intestinal secretion and cyclic AMP production by somatostatin in the rat. Gastroenterology 74:726–730

Cartwright C, McRoberts J, Masui H, Lindeborg D, Dharmsathaphorn K (1984) Quinidine reversal of VIP-stimulated chloride secretion in a human colonic epithelial cell line. Gastroenterology 86:1042

Cassuto J, Fahrenkrug J, Jodal M, Tuttle R, Lundgren O (1981 a) Release of vasoactive intestinal polypeptide from the cat small intestine exposed to cholera toxin. Gut 22:958–963

Cassuto J, Jodal M, Tuttle R, Lundgren O (1981 b) On the role of intramural nerves in the pathogenesis of cholera toxin-induced intestinal secretion. Scand J Gastroenterol 16:377–384

Cassuto J, Jodal M, Tuttle R, Lundgren O (1982 a) 5-hydroxytryptamine and cholera secretion: physiological and pharmacological studies in cats and rats. Scand J Gastroenterol 17:695–703

Cassuto J, Sjövall H, Jodal M, Svanik J, Lundgren O (1982 b) The adrenergic influence on intestinal secretion in cholera. Acta Physiol Scand 115:157–158

Cassuto J, Siewert A, Jodal M, Lundgren O (1983) The involvement of intramural nerves in cholera toxin induced intestinal secretion. Acta Physiol Scand 117:195–202

Cavazzana P, Borsetto PL (1948) Recherches sur l'aspect microscopique des plexus nerveux intramuraux et sur les modifications morphologiques des leurs neurones dans les divers traits de l'intestin humain pendant la vie. Acta Anat (Basel) 5:17–41

Chang EB, Field M, Miller RJ (1982) a_2-Adrenergic receptor regulation of ion transport in rabbit ileum. Am J Physiol 242:G237–G242

Chang EB, Field M, Miller RJ (1983) Enterocyte a_2-adrenergic receptors: yohimbine and p-aminoclonidine binding relative to ion transport. Am J Physiol 244:G76–G82

Chang EB, Brown DR, Field M, Miller RJ (1984) An anti-absorptive basis for precipitated withdrawal diarrhea in morphine-dependent rats. J Pharmacol Exp Ther 228:364–369

Chang EB, Bergenstal RM, Field M (1985) Diarrhea in streptozocin-treated rats. Loss of adrenergic regulation of intestinal fluid and electrolyte transport. J Clin Invest 75:1666–1670

Clague JR, Sternini C, Brecha NC (1985) Localization of calcitonin gene-related peptide-like immunoreactivity in neurons of the rat gastrointestinal tract. Neurosci Lett 56:63–68

Collin J, Kelly KA, Phillips SF (1979) Enhancement of absorption from the intact and transected canine small intestine by electrical pacing. Gastroenterology 76:1422–1428

Conley D, Coyne MJ, Chung A, Bonorris GG, Schoenfield LJ (1976) Propranolol inhibits adenylate cyclase and secretion stimulated by deoxycholic acid in the rabbit colon. Gastroenterology 71:72–75

Cooke HJ (1984) Influence of enteric cholinergic neurons on mucosal transport in guinea-pig ileum. Am J Physiol 246:G263–G267

Cooke HJ (1986) Neurobiology of the intestinal mucosa. Gastroenterology 90:1057–1081

Cooke HJ, Carey HV (1984) The role of enteric nerves in cholera toxin-induced intestinal secretion in the guinea-pig ileum. Gastroenterology 86:1053

Cooke HJ, Carey HV (1985) Pharmacological analysis of 5-hydroxytryptamine actions on guinea-pig ileal mucosa. Eur J Pharmacol 111:329–337

Cooke HJ, Shonnard K, Highison G, Wood JD (1983a) Effects of neurotransmitter release on mucosal transport in guinea-pig ileum. Am J Physiol 245:G745–G750

Cooke HJ, Shonnard K, Wood JD (1983b) Effects of neuronal stimulation on mucosal transport in guinea-pig ileum. Am J Physiol 245:G290–G296

Costa M, Furness JB (1983) The origins, pathways and terminations of neurons with VIP-like immunoreactivity in the guinea-pig small intestine. Neuroscience 8:665–676

Costa M, Furness JB (1984) Somatostatin is present in a subpopulation of noradrenergic nerve fibres supplying the intestine. Neuroscience 13:911–920

Costa M, Gabella G (1971) Adrenergic innervation of the alimentary canal. Z Zellforsch 122:357–377

Costa M, Furness JB, Gabella G (1971) Catecholamine containing nerve cells in the mammalian myenteric plexus. Histochemie 25:103–106

Costa M, Furness JB, Llewellyn-Smith IJ, Davies B, Oliver J (1980) An immunohistochemical study of the projections of somatostatin-containing neurons in the guinea-pig intestine. Neuroscience 5:841–852

Costa M, Furness JB, Llewellyn-Smith IJ, Cuello AC (1981) Projections of substance P-containing neurons within the guinea-pig small intestine. Neuroscience 6:411–424

Costa M, Furness JB, Yanaihara N, Moody TW (1984) Distribution and projections of neurons with immunoreactivity for both gastrin-releasing peptide and bombesin in the guinea-pig small intestine. Cell Tissue Res 235:285–293

Cotterell BJ, Parsons BJ, Poat JA, Roberts PA (1983) A study of rat jejunal a-adrenoceptors. Br J Pharmacol 78:73P

Coupar IM (1976) Stimulation of sodium and water secretion without inhibition of glucose absorption in the rat jejunum by vasoactive intestinal peptide (VIP). Clin Exp Pharmacol Physiol 3:615–618

Coupar IM (1978) Inhibition by morphine of prostaglandin-stimulated fluid secretion in rat jejunum. Br J Pharmacol 63:57–63

Coupar IM (1983) Characterization of the opiate receptor population mediating inhibition of VIP-induced secretion from the small intestine of the rat. Br J Pharmacol 80:371–376

Couvineau A, Royer-Fessard C, Fournier A, St. Pierre S, Pipkorn R, Laburthe M (1984) Structural requirements for VIP interaction with specific receptors in human and rat intestinal membranes: effect of nine partial sequences. Biochem Biophys Res Commun 121:493–498

Coyne MJ, Bonorris GG, Chung A, Conley D, Schoenfield LJ (1977) Propranolol inhibits bile acid and fatty acid stimulation of cyclic AMP in human colon. Gastroenterology 73:971–974

Cuthbert AW, Hickman ME (1985) Indirect effects of adenosine triphosphate on chloride secretion in mammalian colon. J Membrane Biol 86:157–166

Dahlström A, Newson B, Naito S, Ueda T, Ahlman A (1984) Further evidence for the presence of sub-epithelial nerve cells in the rat ileum – an immunohistochemical study. Acta Physiol Scand 120:1–6

Daniel EE, Costa M, Furness JB, Keast JR (1985) Peptide neurons in the canine small intestine. J Comp Neurol 237:227–238

Daniel EE, Furness JB, Costa M, Belbeck L (1987) The projections of chemically identified nerve fibres in canine ileum. Cell Tissue Res 247:377–384

Daumerie J, Henquin JC (1982) Somatostatin and the intestinal transport of glucose and other nutrients in the anaesthetised rat. Gut 23:140–145

Davis GR, Camp RC, Raskin P, Krejs GJ (1980) Effect of somatostatin infusion on jejunal water and electrolyte transport in a patient with secretory diarrhea due to malignant carcinoid syndrome. Gastroenterology 78:346–349

Dermietzel R (1971) Elektronenmikroskopische Untersuchung über die Innervation der Pars pylorica des Mäusemagens. Z Mikrosk Anat Forsch 84:225–256

Desaki J, Fujiwara T, Komuro T (1984) A cellular reticulum of fibroblast-like cells in the rat intestine. Scanning and transmission electron microscopy. Arch Histol Jpn 47:179–186

Dharmsathaphorn K, Pandol SJ (1986) Mechanism of chloride secretion induced by carbachol in a colonic epithelial cell line. J Clin Invest 77:348–354

Dharmsathaphorn K, Binder HJ, Dobbins JW (1980a) Somatostatin stimulates sodium and chloride absorption in the rabbit ileum. Gastroenterology 78:1559–1564

Dharmsathaphorn K, Sherwin RS, Dobbins JW (1980b) Somatostatin inhibits fluid secretion in the rat jejunum. Gastroenterology 78:1554–1558

Dharmsathaphorn K, Harms V, Yamashiro DJ, Hughes RJ, Binder HJ, Wright EM (1983) Preferential binding of vasoactive intestinal polypeptide to basolateral membrane of rat and rabbit enterocytes. J Clin Invest 71:27–35

Dharmsathaphorn K, McRoberts JA, Mandel JG, Tisdale LD, Masui H (1984) A human colonic tumor cell line that maintains vectorial electrolyte transport. Am J Physiol 246:G204–G208

Dharmsathaphorn K, Mandel KG, Masui H, McRoberts JA (1985) Vasoactive intestinal polypeptide-induced chloride secretion by a colonic epithelial cell line. Direct participation of a basolaterally localized Na^+, K^+, Cl^- cotransport system. J Clin Invest 75:462–471

Dimaline R, Dockray GJ (1978) Multiple immunoreactive forms of vasoactive intestinal peptide in human colonic mucosa. Gastroenterology 75:387–392

Dimaline R, Vaillant C, Dockray GJ (1980) The use of region-specific antibodies in the characterization and localization of vasoactive intestinal polypeptide-like substance in the rat gastrointestinal tract. Regul Pept 1:1–16

Dobbins JW, Racusen L, Binder HJ (1980) Effect of D-alanine methionine-enkephaline amide on ion transport in rabbit ileum. J Clin Invest 66:19–28

Dobbins JW, Dharmsathaphorn K, Racusen L, Binder HJ (1981) The effect of somatostatin and enkephalin in ion transport in the intestine. Ann NY Acad Sci 372:594–612

Dockray GJ, Vaillant C, Walsh JH (1979) The neuronal origin of bombesin-like immunoreactivity in the rat gastrointestinal tract. Neuroscience 4:1561–1568

Dogiel AS (1896) Zwei Arten sympathischer Nervenzellen. Anat Anz 11:679–687

Doherty NS, Hancock AA (1983) Role of alpha-2 adrenergic receptors in the control of diarrhea and intestinal motility. J Pharmacol Exp Ther 225:269–274

Donowitz M, Charney AN (1979) Propranolol prevention of cholera enterotoxin-induced secretion in the rat. Gastroenterology 76:482–491

Donowitz M, Fogel R, Battisti L, Asarkof N (1982) The neurohumoral secretagogues carbachol, substance P and neurotensin increase calcium ion flux and calcium content in rabbit ileum. Life Sci 31:1929–1937

Drasch O (1881) Beiträge zur Kenntnis des feineren Baues des Dünndarms insbesondere über die Nerven desselben. Sitzungsber Akad Wiss Wien Math Natur 82:168–198

Driel C, Drukker J van (1973) A contribution to the study of the architecture of the autonomic nervous system of the digestive tract of the rat. J Neural Transm 34:301–320

Dupont C, Laburthe M, Broyart JP, Bataille D, Rosselin G (1980) Cyclic AMP production in isolated colonic epithelial crypts: a highly sensitive model for the evaluation of vasoactive intestinal peptide in human intestine. Eur J Clin Invest 10:67–76

Durbin T, Rosenthal L, McArthur K, Anderson D, Dharmsathaphorn K (1982) Clonidine and lidamidine (WHR-1142) stimulate sodium and chloride absorption in the rabbit intestine. Gastroenterology 82:1352–1358

Ekblad E, Ekelund M, Graffner H, Håkanson R, Sundler F (1985) Peptide-containing nerve fibres in the stomach wall of rat and mouse. Gastroenterology 89:73–85

Eklund S, Jodal M, Lundgren O, Sjöqvist A (1979) Effects of vasoactive intestinal polypeptide on blood flow, motility and fluid transport in the gastrointestinal tract of the cat. Acta Physiol Scand 105:461–468

Eklund S, Cassuto J, Jodal M, Lundgren O (1984) The involvement of the enteric nervous system in the intestinal secretion evoked by cyclic adenosine 3',5'-monophosphate. Acta Physiol Scand 120:311–316

Eklund S, Jodal M, Lundgren O (1985) The enteric nervous system participates in the secretory response to the heat stable enterotoxins of *Escherichia coli* in rats and cats. Neuroscience 14:673–681

Emson PC, de Quidt ME (1984) NPY – a new member of the pancreatic polypeptide family. TINS 7:31–34

Farack JM, Loeschke K (1984) Inhibition by loperamide of deoxycholic acid induced intestinal secretion. Naunyn-Schmiedeberg's Arch Pharmacol 325:286–289

Farack UM, Kautz U, Loeschke K (1981) Loperamide reduces the intestinal secretion but not the mucosal cyclic AMP accumulation induced by cholera toxin. Naunyn-Schmiedeberg's Arch Pharmacol 317:178–179

Favus MJ, Berelowitz M, Coe FL (1981) Effects of somatostatin on intestinal calcium transport in the rat. Am J Physiol 241:215–221

Fehér E, Léránth C (1983) Light and electron microscopic immunocytochemical localization of vasoactive intestinal polypeptide VIP-like activity in the rat small intestine. Neuroscience 10:97–106

Fehér E, Vajda J (1974) Degeneration analysis of the extrinsic nerve elements of the small intestine. Acta Anat (Basel) 87:97–109

Fehér E, Wenger T (1981) Ultrastructural immunocytochemical localization of substance P in the cat small intestine. Acta Histochem (Jena) 150:137–143

Ferri G-L, Botti PL, Vezzadini P, Biliotti G, Bloom SR, Polak JM (1982) Peptide-containing innervation of the human intestinal mucosa. An immunocytochemical study on whole mount preparations. Histochemistry 76:413–420

Ferri G-L, Adrian TE, Ghatei MA, O'Shaughnessy DJ, Probert L, Lee YC, Buchan AMJ, Polak JM, Bloom SR (1983) Tissue localization and relative distribution of regulatory peptides in separated layers from the human bowel. Gastroenterology 84:777–786

Ferri G-L, Botti P, Biliotti G, Rebecchi L, Bloom SR, Tonelli L, Labo G, Polak JM (1984) VIP-, substance P- and met-enkephalin-immunoreactive innervation of the human gastroduodenal mucosa and Brunner's glands. Gut 25:948–952

Field M, McColl I (1973) Ion transport in rabbit ileal mucosa. III. Effects of catecholamines. Am J Physiol 225:852–857

Field M, Sheerin HE, Henderson A, Smith PL (1975) Catecholamine effects on cyclic AMP levels and ion secretion in rabbit ileal mucosa. Am J Physiol 229:86–92

Flemström G, Heylings JR, Garner A (1982) Gastric and duodenal HCO_3^--transport in vitro: effects of hormones and local transmitters. Am J Physiol 242:G100–G110

Flemström G, Kivilaakso E, Briden S, Nylander O, Jedstedt G (1985) Gastroduodenal bicarbonate secretion in mucosal protection. Possible role of vasoactive intestinal peptide and opiates. Dig Dis Sci 30:63–68

Florey HW, Wright RD, Jennings MA (1941) The secretions of the intestine. Physiol Rev 21:36–69

Fogel R, Kaplan RB (1984) Role of enkephalins in regulation of basal intestinal water and ion absorption in the rat. Am J Physiol 246:G386–G392

Freedman J, Rasmussen H, Dobbins JW (1980) Somatostatin stimulates coupled sodium chloride influx across the brush border of the rabbit ileum. Biochem Biophys Res Commun 97:243–247

Friel DD, Miller RJ, Walker MW (1986) Neuropeptide Y: a powerful modulator of epithelial ion transport. Br J Pharmacol 88:425–431

Furness JB, Costa M (1978) Distribution of intrinsic nerve cell bodies and axons which take up aromatic amines and their precursors in the small intestine of the guinea-pig. Cell Tissue Res 188:527–543

Furness JB, Costa M (1987) The enteric nervous system. Churchill Livingstone, Edinburgh

Furness JB, Eskay RL, Brownstein MJ, Costa M (1980) Characterization of somatostatin-like immunoreactivity in intestinal nerves by high pressure liquid chromatography and radioimmunoassay. Neuropeptides 1:97–103

Furness JB, Costa M, Eckenstein F (1983a) Neurons localized with antibodies against choline acetyltransferase in the enteric nervous system. Neurosci Lett 40:105–109

Furness JB, Costa M, Emson PC, Håkanson R, Moghimzadeh E, Sundler F, Taylor IL, Chance RE (1983b) Distribution, pathways and reactions to drug treatment of nerves with neuropeptide Y and pancreatic polypeptide-like immunoreactivity in the guinea-pig digestive tract. Cell Tissue Res 234:71–92

Furness JB, Costa M, Miller RJ (1983c) Distribution and projections of nerves with enkephalin-like immunoreactivity in the guinea-pig small intestine. Neuroscience 8:653–664

Furness JB, Costa M, Keast JR (1984) Choline acetyltransferase and peptide immunoreactivity of submucous neurons in the small intestine of the guinea-pig. Cell Tissue Res 237:328–336

Furness JB, Costa M, Gibbins IL, Llewellyn-Smith IJ, Oliver JR (1985) Neurochemically similar myenteric and submucous neurons directly traced to the mucosa of the small intestine. Cell Tissue Res 241:155–163

Furness JB, Llewellyn-Smith IJ, Bornstein JC, Costa M (1986) Neuronal circuitry in the enteric nervous system. In: Owman C, Bjorklund A, Hökfelt T (eds) Handbook of chemical neuroanatomy. Elsevier, Amsterdam

Fuxe K, Hökfelt T, Said S, Mutt V (1977) Vasoactive intestinal polypeptide and the nervous system: immunohistochemical evidence for localization in central and peripheral neurons, particularly intra-cortical neurons of the cerebral cortex. Neurosci Lett 5:241–246

Gabella G, Costa M (1967) Le fibre adrénergiche del canale alimentare. Giorn Acc Med Torino 130:1–12

Gabella G, Costa M (1968) Adrenergic fibres in the mucous membrane of guinea-pig alimentary tract. Experientia 24:706–707

Gabella G, Juorio AV (1975) Effect of extrinsic denervation on endogenous noradrenaline and ^{3}H-[noradrenaline] uptake in the guinea-pig colon. J Neurochem 25:631–634

Gaginella TS (1984) Cholinergic modulation of transport. Fed Proc 43:2930–2931

Gaginella TS, Mekhjian HS, O'Dorisio TM (1978a) Vasoactive intestinal peptide: quantification by radioimmunoassay in isolated cells, mucosa and muscle of the hamster intestine. Gastroenterology 74:718–721

Gaginella TS, O'Dorisio TM, Wu ZC, Mekhjian HS, Cataland S (1978b) Pancreatic polypeptide: effect on fluid transport in the small and large intestine of the rat. Clin Res 26:661

Gaginella TS, O'Dorisio TM, Hubel KA (1981) Release of vasoactive intestinal polypeptide by electrical field stimulation of rabbit ileum. Regul Pept 2:165–174

Gaginella TS, Rimele TJ, Wietecha M (1983) Studies on rat intestinal epithelial cell receptors for serotonin and opiates. J Physiol (Lond) 335:101–111

Gamse R, Saria A, Bucsics A, Lembeck F (1981) Substance P in tumors: pheochromocytoma and carcinoid. Peptides 2 [Suppl 2]:275–280

Ghiglione M, Christofides ND, Yiangou Y, Uttenthal LO, Bloom SR (1982) PHI stimulates intestinal fluid secretion. Neuropeptides 3:79–82

Gibbins IL (1982) Lack of correlation between ultrastructural and pharmacological types of non-adrenergic autonomic nerves. Cell Tissue Res 221:551–581

Goniaew K (1875) Die Nerven des Nahrungsschlauches. Eine histologische Studie. Arch Mikrosk Anat 11:479–496

Gordon SJ, Kinsey MD, Magen JS, Joseph RE, Kowlessar OD (1978) Effect of loperamide on bile acid induced secretion in the rat cecum. Gastroenterology 74:1040–1046

Granger DN, Cross R, Barrowman JA (1982) Effects of various secretagogues and human carcinoid serum on lymph flow in the cat ileum. Gastroenterology 83:896–901

Greenwood B, Davison JS (1985) Role of extrinsic and intrinsic nerves in the relationship between intestinal motility and transmural potential difference in the anaesthetized ferret. Gastroenterology 89:1286–1292

Guandalini S, Kachur JF, Smith PL, Miller RJ, Field M (1980) In vitro effects of somatostatin on ion transport in rabbit intestine. Am J Physiol 283:G67–G74

Hanau A (1886) Experimentelle Untersuchungen über die Physiologie der Darmsekretion. Z Biol 22:195–235

Hansson H-A, Lange S, Lönnroth I (1984) Internalization in vivo of cholera toxin in the small intestinal epithelium of the rat. Acta Pathol Microbiol Immunol Scand [B] 92:15–21

Hardcastle PT, Eggenton J (1973) The effect of acetylcholine on the electrical activity of intestinal epithelial cells. Biochim Biophys Acta 298:95–100

Hardcastle J, Hardcastle PT, Read NW, Redfern JS (1981a) The action of loperamide in inhibiting prostaglandin induced intestinal secretion in the rat. Br J Pharmacol 74:563–569

Hardcastle J, Hardcastle PT, Redfern JS (1981b) Action of 5-hydroxytryptamine on intestinal ion transport in the rat. J Physiol (Lond) 320:41–55

Hardcastle J, Hardcastle PT, Noble JM (1983) The role of calcium in intestinal secretion in the rat in vitro. J Physiol 338:51P

Heitz P, Polak JM, Timson CM, Pearse AGE (1976) Enterochromaffin cells as the endocrine source of gastrointestinal substance P. Histochemistry 49:343–347

Hill CJ (1927) VIII. A contribution to our knowledge of the enteric plexuses. Philos Trans Roy Soc Lond [B] 215:355–387

Hirst GDS, McKirdy HC (1975) Synaptic potentials recorded from neurones of the submucous plexus of guinea-pig small intestine. J Physiol (Lond) 249:369–386

Hökfelt T, Johansson O, Efendic S, Luft R, Arimura A (1975) Are there somatostatin-containing nerves in the rat gut? Immunohistochemical evidence for a new type of peripheral nerves. Experientia 31:852–854

Holzer P, Bucsics A, Saria A, Lembeck F (1982) A study of the concentrations of substance P and neurotensin in the gastrointestinal tract of various mammals. Neuroscience 7:2919–2924

Honjin R (1951) Studies on the nerve endings in the small intestine. Cyt Neurol Stud 9:1–14

Honjin R, Takahashi A (1966) Electron microscopy of synaptic nerve endings in the walls of the digestive tract. Symp Cell Chem 16:59–74

Honjin R, Takahashi A, Tasaki Y (1965) Electron microscopy of nerve endings in the mucous membrane of human intestine. Okajimas Folia Anat Jpn 40:409–427

Hoyes AD, Barber P (1981) Degeneration of axons in the ureteric and duodenal nerve plexuses of the adult rat following in vivo treatment with capsaicin. Neurosci Lett 25:19–24

Hubel KA (1976) Intestinal ion transport: effect of norepinephrine, pilocarpine and atropine. Am J Physiol 231:252–257

Hubel KA (1977) Effects of bethanechol on intestinal ion transport in the rat. Proc Soc Exp Biol Med 154:41–44

Hubel KA (1978) The effects of electrical field stimulation and tetrodotoxin on ion transport by the isolated rabbit ileum. J Clin Invest 62:1039–1047

Hubel KA (1983) Effects of scorpion venom on electrolyte transport by rabbit ileum. Am J Physiol 244:G501–G506

Hubel KA (1984) Electrical stimulus-secretion coupling in rabbit ileal mucosa. J Pharmacol Exp Ther 231:577–582

Hubel KA (1985) Intestinal nerves and ion transport: stimuli, reflexes, and responses. Am J Physiol 248:G261–G271

Hubel KA, Callanan D (1980) Effects of calcium ions on ileal transport and electrically induced secretion. Am J Physiol 239:G18–G22

Hubel KA, Renquist KS (1985) Neuropeptide Y selectively increases chloride absorption. Gastroenterology 88:1423

Hubel KA, Shirazi S (1982) Human ileal transport in vitro: changes with electrical field stimulation and tetrodotoxin. Gastroenterology 83:63–68

Hubel KA, Renquist KS, Shirazi S (1983) Intramural cholinergic nerves affect mucosal ion transport by the left colon of man. Gastroenterology 64:1192

Hubel KA, Renquist KS, Shirazi S (1984) Neural control of ileal ion transport: role of substance P (SP) in rabbit and man. Gastroenterology 86:1118

Hughes S, Higgs NB, Turnberg LA (1982) Antidiarrhoeal activity of loperamide: studies of its influence on ion transport across rabbit ileal mucosa in vitro. Gut 23:974–979

Hughes S, Higgs NB, Turnberg LA (1984) Loperamide has anti-secretory activity in the human jejunum in vivo. Gut 25:931–935

Hukuhara T, Yamagami M, Nakayama S (1958) On the intestinal intrinsic reflexes. Jpn J Physiol 8:9–20

Isaacs PET, Corbett CL, Riley AK, Hawker PC, Turnberg LA (1976) In vitro behaviour of human intestinal mucosa: the influence of acetylcholine on ion transport. J Clin Invest 58:535–542

Isaacs PET, Whitehead JS, Kim YS (1982) Muscarinic acetylcholine receptors of the small intestine and pancreas of the rat: distribution and the effect of vagotomy. Clin Sci 62:203–207

Isenberg JI, Wallin B, Johanssen C, Smedfors B, Mutt V, Tatemoto K, Emas S (1984) Secretin, VIP and PHI stimulate rat proximal duodenal surface epithelial bicarbonate secretion in vivo. Regul Pept 8:315–320

Ishikawa N (1926) Experimentelle Untersuchungen über die Dickdarminnervation, insbesondere des Colon descendens und sigmoideum. Jpn J Med Sci 3:21–22

Ito T, Kubo M (1940) Zytologische Untersuchungen über die intramuralen Ganglienzellen des Verdauungstraktes. Über die Ganglienzellen des menschlichen Darmes, mit besonderer Berücksichtigung auf die Nisslsubstanz. Cytologia (Tokyo) 10:334–347

Ito S, Iwanaga T, Yamada Y, Shibata A (1982) Somatostatin-28 like immunoreactivity in the human gut. Horm Metab Res 14:500–501

Jacobowitz D (1965) Histochemical studies of the autonomic innervation of the gut. J Pharmacol Exp Ther 149:358–364

Jaros W, Biller J, Greer S, Grand R (1985) Successful treatment of idiopathic secretory diarrhea of infancy with a somatostatin analogue SMS 201-995. Gastroenterology 88:1432

Jessen KR, Saffrey MJ, van Noorden S, Bloom SR, Polak JM, Burnstock G (1980) Immunohistochemical studies of the enteric nervous system in tissue culture and in situ: localization of vasoactive intestinal polypeptide (VIP), substance P and enkephalin immunoreactive nerve in the guinea-pig gut. Neuroscience 5:1717–1735

Kachur JF, Miller RJ (1982) Characterization of the opiate receptor in the guinea-pig ileal mucosa. Eur J Pharmacol 81:177–183

Kachur JF, Miller RJ, Field M (1980) Control of guinea pig intestinal electrolyte secretion by a delta-opiate receptor. Proc Natl Acad Sci USA 77:2753–2756

Kachur JF, Miller RJ, Field M, Rivier J (1982) Neurohumoral control of ileal electrolyte transport. II. Neurotensin and substance P. J Pharmacol Exp Ther 20:456–463

Karlström L, Cassuto J, Jodal M, Lundgren O (1983) The importance of the enteric nervous system for the bile salt-induced secretion in the small intestine of the rat. Scand J Gastroenterol 18:117–123

Keast JR, Furness JB, Costa M (1984a) Somatostatin in human enteric nerves. Distribution and characterization. Cell Tissue Res 237:299–308

Keast JR, Furness JB, Costa M (1984b) The origins of peptide and norepinephrine nerves in the mucosa of the guinea pig small intestine. Gastroenterology 86:637–644

Keast JR, Furness JB, Costa M (1985a) Different substance P receptors are found on mucosal epithelial cells and submucous neurons of the guinea-pig small intestine. Naunyn-Schmiedeberg's Arch Pharmacol 329:382–387

Keast JR, Furness JB, Costa M (1985b) Distribution of certain peptide-containing nerves and endocrine cells in the gastrointestinal mucosa in five mammalian species. J Comp Neurol 236:403–422

Keast JR, Furness JB, Costa M (1985 c) Investigations of nerve populations influencing ion transport that can be stimulated electrically, by serotonin and by a nicotinic agonist. Naunyn-Schmiedeberg's Arch Pharmacol 331:260–266

Keast JR, Furness JB, Costa M (1986) Effects of noradrenaline and somatostatin on basal and stimulated mucosal ion transport in the guinea-pig small intestine. Naunyn-Schmiedeberg's Arch Pharmacol 337:393–399

Keast JR, Furness JB, Costa M (1987) Distribution of peptide-containing neurons and endocrine cells in the rabbit gastrointestinal tract, with particular reference to the mucosa. Cell Tissue Res 248:565–567

Kishimoto S, Konemori R, Mikai T, Kambara A, Okamotu K, Shimizu S, Iwasaki T, Daitoku K, Kajiyama G, Miyoshi A, Yanaihara N (1984) VIPergic innervation in rats with congenital aganglionic colon. Hiroshima J Med Sci 33:369–376

Klaeveman HL, Conlon TP, Levy AG, Gardner JD (1975) Effects of gastrointestinal hormones on adenylate cyclase activity in human jejunal mucosa. Gastroenterology 68:667–675

Knaffl-Lenz G, Nagaki S (1925) Über die Resorption aus ausgeschalteten Darmschlingen. Naunyn-Schmiedeberg's Arch Exp Path Pharmak 105:109–123

Kobayashi S, Suzuki M, Uchida T, Yanaihara N (1984) Enkephalin neurons in the guinea pig duodenum: a light and electron microscopic immunocytochemical study using an antiserum to Methionine-enkephalin-Arg^6-Gly^7-Leu^8. Biomed Res 5:489–506

Krasny EJ, Frizzell RA (1984) Fluid secretion by isolated colonic crypts. Fedn Proc 43:1087

Krause (1861) Anatomische Untersuchungen. Hannover p 641 (cited from Breiter and Frey, 1862)

Krejs GJ (1984) Effect of somatostatin infusion on VIP-induced transport changes in the human jejunum. Peptides 5:271–276

Krejs GJ, Fordtran JS (1980) Effect of VIP infusion on water and ion transport in the human jejunum. Gastroenterology 78:722–777

Krejs GJ, Barkley RM, Read NW, Fordtran JS (1978) Intestinal secretion induced by VIP: a comparison with cholera toxin in canine jejunum in vivo. J Clin Invest 61:1337–1345

Krejs GJ, Browne R, Raskin P (1980) Effect of intravenous somatostatin on jejunal absorption of glucose, amino acids, water and electrolytes. Gastroenterology 78:26–31

Krokhina EM (1973) The sympathetic innervation of the gastrointestinal tract of mammals. Arch Anat Micr 62:307–321

Kuntz A (1913) On the innervation of the digestive tube. J Comp Neurol 23:173–192

Kuntz A (1922) On the occurrence of reflex arcs in the myenteric and submucous plexuses. Anat Rec 24:193–210

Kvietys PR, Granger DN, Mortillaro NA, Taylor AE (1979) Effect of atropine on fatty acid induced changes in jejunal blood flow, oxygen consumption and water transport. Physiologist 22:74

Laburthe M, Prieto JC, Amiranoff B, Dupont C, Hui Bon Hoa D, Rosselin G (1979) Interaction of vasoactive intestinal peptide with isolated intestinal epithelial cells from rat. 2. Characterization and structural requirements of the stimulatory effect of vasoactive intestinal peptide on production of adenosine 3′,5′-monophosphate. Eur J Biochem 96:239–248

Laburthe M, Couvineau A, Rouyer-Fessard C, Moroder L (1985) Interaction of PHM, PHI and 24-glutamine PHI with human VIP receptors from colonic epithelium: comparison with rat intestinal receptors. Life Sci 36:991–995

Larsson L-I (1977) Ultrastructural localization of a new neuronal peptide (VIP). Histochemistry 54:173–176

Larsson L-I, Fahrenkrug J, Schaffalitzky de Muckadell OB, Sundler F, Håkanson R, Rehfeld JF (1976) Localization of vasoactive intestinal polypeptide (VIP) to central and peripheral neurons. Proc Natl Acad Sci USA 73:3197–3200

Larsson L-I, Polak JM, Buffa R, Sundler F, Solcia E (1979) On the immunocytochemical localization of the vasoactive intestinal polypeptide. J Histochem Cytochem 27:936–938

Lassmann G (1975) Vorkommen von Ganglienzellen im Schleimhautstroma von Colon, Sigma und Rectum. Virchows Arch [A] 365:257–261

Leander S, Ekman R, Uddman R, Sundler F, Håkanson R (1984) Neuronal cholecystokinin, gastrin-releasing peptide, neurotensin, and β-endorphin in the intestine of the guinea pig. Distribution and possible motor functions. Cell Tissue Res 235:521−531

Levens NR (1985) Control of intestinal absorption by the renin-angiotensin system. Am J Physiol 249:G3−G15

Llewellyn-Smith IJ, Furness JB, Murphy R, O'Brien PE, Costa M (1984a) Substance P-containing nerves in the human small intestine. Distribution, ultrastructure and characterization of the immunoreactive peptide. Gastroenterology 86:421−435

Llewellyn-Smith IJ, Furness JB, O'Brien PE, Costa M (1984b) Noradrenergic nerves in human small intestine. Distribution and ultrastructure. Gastroenterology 87:513−529

Lolova I, Itzev D, Davidoff M (1984) Immunocytochemical localization of substance P, methionine-enkephalin and somatostatin in the cat intestinal wall. J Neural Transmission 60:71−88

Lönnroth I, Jennische E (1982) Reversal of enterotoxic diarrhoea by anaesthetic and membrane-stabilizing agents. Acta Pharmacol Toxicol (Copenh) 51:330−335

Lopéz-Ruiz MP, Arilla E, Goméz-Pan A, Prieto JC (1985) Interaction of Leu-enkephalin with isolated enterocytes from guinea-pig: binding to specific receptors and stimulation of cAMP accumulation. Biochem Biophys Res Commun 126:404−411

Lorén I, Alumets J, Håkanson R, Sundler F (1979) Immunoreactive pancreatic polypeptide (PP) occurs in the central and peripheral nervous system: preliminary immunocytochemical observations. Cell Tissue Res 200:179−186

Lundberg JM, Dahlström A, Bylock A, Ahlman H, Petterson G, Larsson I, Hansson H-A, Kewenter J (1978) Ultrastructural evidence for an innervation of epithelial enterochromaffine cells in the guinea pig duodenum. Acta Physiol Scand 104:3−12

Lundberg JM, Terenius L, Hökfelt T, Martling CR, Tatemoto K, Mutt V, Polak J, Bloom S, Goldstein M (1982) Neuropeptide Y (NPY)-like immunoreactivity in peripheral noradrenergic neurons and effects of NPY on sympathetic function. Acta Physiol Scand 116:477−480

McFadden D, Zinner MJ, Jaffe BM (1986) Substance P-induced intestinal secretion of water and electrolytes. Gut 27:267−272

McFadyen RJ, Allen JM, Bloom SR (1986) NPY stimulates net absorption across rat intestinal mucosa in vivo. Neuropeptides 7:219−227

McKay JS, Linaker BD, Turnberg LA (1981) The influence of opiates on ion transport across rabbit ileal mucosa. Gastroenterology 80:279−284

Macrae IM, Furness JB, Costa M (1986) Distribution of subgroups of noradrenaline neurons in the coeliac ganglion of the guinea-pig. Cell Tissue Res 244:173−180

Maeda M, Takagi H, Kubota Y, Morishima Y, Akai F, Hashimoto S, Mori S (1985) The synaptic relationship between vasoactive intestinal polypeptide (VIP)-like immunoreactive neurons and their axon terminals in the rat small intestine: light and electron microscopic study. Brain Res 329:356−359

Mailman D (1978) Effects of VIP on intestinal absorption and blood flow. J Physiol (Lond) 297:121−132

Mailman D (1980) Effects of morphine on canine intestinal absorption and blood flow. Br J Pharmacol 68:617−624

Mailman D (1984a) Morphine-neural interactions on canine intestinal absorption and blood flow. Br J Pharmacol 81:263−270

Mailman D (1984b) Effects of atropine and guanethidine on canine intestinal absorption and blood flow. Life Sci 34:1309−1315

Makino K (1955) A histological study of sensory nerves in the small intestine and cecum. Arch Jpn Chir 24:443−455

Malmfors G, Leander S, Brodin E, Håkanson R, Holmin T, Sundler F (1981) Peptide-containing neurons intrinsic to the gut wall. An experimental study in the pig. Cell Tissue Res 214:225−238

Märki F (1981) Effect of somatostatin on intestinal absorption of nutrients in the rat. Regul Pept 2:371−381

Matthews MR, Cuello AC (1984) The origin and possible significance of substance P immunoreactive networks in the prevertebral ganglia and related structures in the guinea-pig. Phil Trans Roy Soc Lond [B] 306:247–276

Mazzanti L, del Tacca M, Breschi MC, Frigo GM, Friedman C, Crema A (1972) The time course of functional and morphological changes of the guinea-pig colon after 'a frigore' denervation of the periarterial sympathetic nerves. Acta Neuropathol (Berl) 22:190–199

Melander T, Hökfelt T, Rökaeus Å, Fahrenkrug J, Tatemoto K, Mutt V (1985) Distribution of galanin-like immunoreactivity in the gastrointestinal tract of several mammalian species. Cell Tissue Res 239:253–270

Mihara S, North RA (1986) Opioids increase potassium conductance in submucous neurones of guinea-pig caecum by activating δ-receptors. Br J Pharmacol 88:315–322

Mitchenere P, Adrian TE, Kirk RM, Bloom SR (1981) Effect of gut regulatory peptides on intestinal luminal fluid in the rat. Life Sci 29:1563–1570

Modlin JM, Bloom SR, Mitchell SJ (1978) Experimental evidence for VIP as a cause of the watery diarrhoea syndrome. Gastroenterology 75:1051–1054

Moghimzadeh E, Ekman R, Håkanson R, Yanaihara N, Sundler F (1983) Neuronal gastrin-releasing peptide in the mammalian gut and pancreas. Neuroscience 10:553–563

Molnár B (1909) Zur Analyse des Erregungs- und Hemmungsmechanismus der Darmsaftsekretion. Dtsch Med Wochenschr 35:1384–1385

Moriarty KJ, Hegarty JE, Tatemoto K, Mutt V, Christofides ND, Bloom SR, Wood JR (1984) Effect of peptide histidine isoleucine on water and electrolyte transport in the human jejunum. Gut 25:624–628

Moriarty KJ, Higgs NB, Woodford M, Turnberg LA (1985) Cholera toxin stimulates secretion in stripped rabbit ileum independently of enteric neural reflexes. Gastroenterology 88:1507

Morris AI, Turnberg LA (1980) The influence of a parasympathetic agonist and antagonist on human intestinal transport in vitro. Gastroenterology 79:861–866

Morris AI, Turnberg LA (1981) Influence of isoproterenol and propranolol in human intestinal transport in vivo. Gastroenterology 81:1076–1079

Müller E (1892) Zur Kenntnis der Ausbreitung und Endigungsweise der Magen-, Darm- und Pankreas-Nerven. Arch Mikr Anat 40:390–409

Müller E (1921) Über das Darmnervensystem. Uppsala Läk För, NF 26:1–22

Müller LR (1911) Die Darminnervation. Dtsch Arch Klin Med 105:1–43

Nakaki T, Nakadate T, Yamamoto S, Kato R (1982) Alpha-2 adrenergic inhibition of intestinal secretion induced by prostaglandin E_1, vasoactive intestinal peptide and dibutyryl cyclic AMP in rat jejunum. J Pharmacol Exp Ther 220:637–641

Nasset ES, Pierce HB, Murlin HR (1935) Proof of a humoral control of intestinal secretion. Am J Physiol 111:145–158

Newson B, Ahlman H, Dahlström A, das Gupta TK, Nyhus LM (1979a) Are there sensory neurons in the mucosa of the mammalian gut? Acta Physiol Scand 105:521–523

Newson B, Ahlman H, Dahlström A, das Gupta TK, Nyhus LM (1979b) On the innervation of the ileal mucosa in the rat – a synapse. Acta Physiol Scand 105:387–389

Newson B, Dahlström A, Enerbäck L, Ahlman H (1983) Suggestive evidence for a direct innervation of mucosal mast cells. An electron microscopic study. Neuroscience 10:565–570

Nilsson G, Larsson L-I, Håkanson R, Brodin E, Pernow B, Sundler F (1975) Localization of substance P-like immunoreactivity in mouse gut. Histochemistry 43:97–99

Nilsson O, Cassuto J, Larsson P-A, Jodal M, Lidberg P, Ahlman H, Dahlström A, Lundgren O (1983) 5-Hydroxytryptamine and cholera secretion: a histochemical and physiological study in cats. Gut 24:542–548

Norberg K-A (1964) Adrenergic innervation of the intestinal wall studied by fluorescence microscopy. Int J Neuropharmacol 3:379–384

North RA, Surprenant A (1985) Inhibitory synaptic potentials resulting from α_2-adrenoceptor activation in guinea-pig submucous plexus neurons. J Physiol (Lond) 358:17–33

Ohkubo K (1936) Studies on the intrinsic nervous system of the digestive tract. I. The submucous plexus of guinea pigs. Jpn J Med Sci 6:1–20

Ohkubo K (1937) Studien über das intramurale Nervensystem des Verdauungskanales. III. Affe und Mensch. Jpn J Med Sci Anat 6:219–247

Okamura C (1929) Zur Vervollkommnung des Nervenapparatus in der Wand des Verdauungstraktes. Z Anat Entwick 91:627–632

Oshima L (1929) Über die Innervation des Darmes. Z Anat Entwick 90:725–767

Palay SL, Karlin LJ (1959) An electron microscope study of the intestinal villus. I. The fasting animal. J Biophys Biochem Cytol 5:363–371

Parsons BJ, Poat JA, Roberts PA (1983) a-Receptors associated with fluid absorption in rat jejunum and ileum. Br J Pharmacol 79:307P

Parsons BJ, Poat JA, Roberts P (1984) Studies of the mechanism of noradrenaline stimulation of fluid absorption by rat jejunum in vitro. J Physiol (Lond) 355:427–439

Pearse AGE, Polak JM (1975) Immunocytochemical localization of substance P in mammalian intestine. Histochemistry 41:373–375

Penman E, Wass JAH, Butler MG, Penny ES, Price J, Wu P, Rees LH (1983) Distribution and characterisation of immunoreactive somatostatin in human gastrointestinal tract. Regul Pept 7:53–65

Pick J (1967) Fine structure of nerve terminals in the human gut. Anat Rec 159:131–146

Pitha J (1969) Early fine structural changes in the rabbit upper ileum after superior mesenteric sympathectomy, with special reference to the mucosa. J Ultrastruct Res 26:529–539

Pott G, Wagner H, Zierden E, Hilke KH, Jansen H, Hengst K, Gerlach V (1979) Influence of somatostatin on carbohydrate absorption in human small intestine. Klin Wochenschr 57:131–133

Prieto JC, Laburthe M, Rosselin G (1979) Interaction of vasoactive intestinal peptide with isolated intestinal epithelial cells from rat. 1. Characterization, quantitative aspects and structural requirements of binding sites. Eur J Biochem 96:229–237

Rabinovitch J (1927) Factors influencing the absorption of water and chlorides from the intestine. Am J Physiol 82:279–289

Racusen LC, Binder HJ (1977) Alteration of large intestinal electrolyte transport by vasoactive intestinal polypeptide in the rat. Gastroenterology 73:790–796

Racusen LC, Binder HJ (1979) Adrenergic interaction with ion transport across colonic mucosa: role of both alpha and beta adrenergic agonists. In: Binder JH (ed) Mechanisms of intestinal secretion. Alan R Liss New York, pp 201–215

Ramón y Cajal S (1894) cited from Ramón y Cajal (1911) In: Maloine A (ed) Histologie du système nerveux de l'homme et des vertébrés. Vol II, pp 891–942

Rangachari PK, McWade D (1986) Epithelial and mucosal preparations of canine proximal colon in Ussing chambers: comparisons of responses. Life Sci 38:1641–1652

Ranson SW (1921) Afferent paths for visceral reflexes. Physiol Rev 1:477–522

Rao MB, O'Dorisio TM, Cataland S, George JM, Gaginella TS (1984) Angiotensin II and norepinephrine antagonize the secretory effect of VIP in rat ileum and colon. Peptides 5:291–294

Read JB, Burnstock G (1968a) Comparative histochemical studies of adrenergic nerves in the enteric plexuses of vertebrate large intestine. Comp Biochem Physiol 27:505–517

Read JB, Burnstock G (1968b) Fluorescent histochemical studies on the mucosa of the vertebrate gastrointestinal tract. Histochemie 16:324–332

Read NW, Smallwood RH, Levin RJ, Holdworth CD, Brown BH (1977) Relationship between changes in intraluminal pressure and transmural potential difference in the human and canine jejunum in vivo. Gut 18:141–151

Redfors S, Sjövall H (1984) The importance of nervous and humoral factors in the control of vascular resistance, blood flow distribution and net fluid absorption in the cat small intestine during hemorrhage. Acta Physiol Scand 121:305–315

Reichert CB (1859) Über die angeblichen Nervenanastomosen im Stratum nerveum S. vasculosum der Darmschleimhaut. Arch Anat Physiol 4:530–536

Reid EW (1892) Preliminary report on experiments upon intestinal absorption without osmosis. Br Med J 1:1133–1134

Reinecke M, Schlüter P, Yanaihara N, Forssman WG (1981) VIP immunoreactivity in enteric nerves and endocrine cells of the vertebrate gut. Peptides 2:149–156

Reiser KA (1933) Über die Endausbreitung des vegetativen Nervensystems. Z Zellforsch 17:610–641

Remak R (1858) Über peripherische Ganglien an den Nerven des Nahrungsrohrs. Arch Anat Physiol Wissenschl Med 2:189–192

Rimele TJ, Gaginella TS (1982) In vivo identification of muscarinic receptors on rat ileal and colonic epithelial cells: binding of ^{3}H-quinuclidinyl benzilate. Naunyn-Schmiedeberg's Arch Pharmacol 319:18–21

Rimele TJ, O'Dorisio MS, Gaginella TS (1981) Evidence for muscarinic receptors on rat colonic epithelial cells: binding of (^{3}H)-quinuclidinyl benzilate. J Pharmacol Exp Ther 218:426–431

Rintoul JR (1960) The comparative morphology of the enteric nerve plexuses. M.D. Thesis, St. Andrews University, pp 1–251

Rosenthal LE, Yamashiro DJ, Rivier J, Vale W, Brown M, Dharmsathaphorn K (1983) Structure-activity relationships of somatostatin analogs in the rabbit ileum and rat colon. J Clin Invest 71:840–849

Sabussow NP (1913) Zur Frage nach der Innervation des Schlundkopfes und der Speiserohre der Säugetiere. Anat Anz 44:64–69

Said SI, Faloona GR (1975) Elevated plasma and tissue levels of vasoactive intestinal polypeptide in the watery-diarrhea syndrome due to pancreatic, bronchogenic and other tumours. N Engl J Med 293:155–160

Sandhu BK, Tripp JH, Candy DCA, Harries JT (1981) Loperamide: studies on its mechanism of action. Gut 22:658–662

Santangelo WC, O'Dorisio TM, Kim JG, Severino G (1985) Effect of a somatostatin analogue on intestinal water and ion transport in pancreatic cholera syndrome. Gastroenterology 88:1570

Saria A, Beubler E (1985) Neuropeptide Y (NPY) and peptide YY (PYY) inhibit prostaglandin E_2-induced intestinal fluid and electrolyte secretion in the rat jejunum in vivo. Eur J Pharmacol 119:47–52

Savitch V, Sochenstvensky NA (1917) L'influence du nerf vague sur la sécrétion de l'intestin. C R Soc Biol (Paris) 69:508–510

Schabadasch A (1930) Intramurale Nervengeflechte des Darmrohrs. Z Zellforsch Mikrosk Anat 10:320–385

Schiller LR, Santa Ana CA, Morawski SG, Fordtran JS (1985) Studies on the antidiarrheal action of clonidine. Effects on motility and intestinal absorption. Gastroenterology 89:982–988

Schofield GC (1960) Experimental studies on the innervation of the mucous membrane of the gut. Brain 83:490–514

Schultzberg M, Hökfelt T, Nilsson G, Terenius L, Rehfeld JF, Brown M, Elde R, Goldstein M, Said S (1980) Distribution of peptide- and catecholamine-containing neurons in the gastrointestinal tract of rat and guinea pig: immunohistochemical studies with antisera to substance P, vasoactive intestinal polypeptide, enkephalins, somatostatin, gastrin/cholecystokinin, neurotensin and dopamine-β-hydroxylase. Neuroscience 5:689–744

Schwartz CJ, Kimberg DV, Sheerin HE, Field M, Said SI (1974) Vasoactive intestinal peptide stimulation of adenylate cyclase and active electrolyte secretion in intestinal mucosa. J Clin Invest 54:536–544

Sellin JH, de Soignie R (1985) Adrenergic regulation of ion transport in rabbit proximal colon. Gastroenterology 88:1580

Silva DG, Ross G, Osborne LW (1971) Adrenergic innervation of the ileum of the cat. Am J Physiol 220:347–352

Simon B, Kather H (1978) Activation of human adenylate cyclase in the upper gastrointestinal tract by vasoactive intestinal polypeptide. Gastroenterology 74:722–725

Simon B, Czygan P, Spaan G, Dittrich J, Kather H (1978) Hormone sensitive adenylate cyclase in human colonic mucosa. Digestion 17:229–233

Sjöqvist A, Cassuto J, Jodal M, Brunsson I, Lundgren O (1982) The effect on intestinal fluid transport of exposing the serosa to hydrochloric acid. A study of mechanisms. Acta Physiol Scand 116:447–454

Sjövall H (1984a) Evidence for separate sympathetic regulation of fluid absorption and blood flow in the feline jejunum. Am J Physiol 247:G510–G514

Sjövall H (1984b) Sympathetic control of jejunal fluid and electrolyte transport. An experimental study in cats and rats. Acta Physiol Scand 122 [Suppl 535]

Sjövall H, Jodal M, Redfors S, Lundgren O (1982) The effect of carotid occlusion on the rate of net fluid absorption in the small intestine of rats and cats. Acta Physiol Scand 115:447–453

Sjövall H, Brunsson I, Jodal M, Lundgren O (1983a) The effect of vagal nerve stimulation on net fluid transport in the small intestine of the cat. Acta Physiol Scand 117:351–357

Sjövall H, Redfors S, Hallbäck D-A, Eklund S, Jodal M, Lundgren O (1983b) The effect of splanchnic nerve stimulation on blood flow distribution, villous tissue osmolality and fluid and electrolyte transport in the small intestine of the cat. Acta Physiol Scand 117:359–365

Sjövall H, Redfors S, Jodal M, Lundgren O (1983c) On the mode of action of the sympathetic fibres on intestinal fluid transport: evidence for the existence of a glucose-stimulated secretory nervous pathway in the intestinal wall. Acta Physiol Scand 119:39–48

Sjövall H, Jodal M, Lundgren O (1984a) Further evidence for a glucose-activated secretory mechanism in the jejunum of the cat. Acta Physiol Scand 120:437–443

Sjövall H, Redfors S, Biber B, Martner J, Winsö O (1984b) Evidence for cardiac volume-receptor regulation of feline jejunal blood flow and fluid transport. Am J Physiol 246:G401–G410

Smith PL, McCabe RD (1984) Effects of adrenergic agents on electrolyte transport by the rabbit descending colon. Fed Proc 43:1082

Smith PL, McCabe RD (1986) Potassium secretion by rabbit descending colon: effects of adrenergic stimuli. Am J Physiol 250:G432–G439

Stach W (1973) Über die Nervengeflechte der Duodenalzotten. Licht- und electronenmikroskopische Untersuchungen. Acta Anat (Basel) 85:216–231

Stach W (1979) Zur Innervation der Dünndarmschleimhaut von Laboratoriumstieren. II. Ultrastruktur der neurozellularen Beziehungen. Z Mikrosk Anat Forsch 93:1012–1024

Stach W, Hung N (1979) Zur Innervation der Dünndarmschleimhaut von Laboratoriumstieren. I. Architektur, lichtmikroskopische Struktur und histochemische Differenzierung. Z Mikrosk Anat Forsch 93:976–987

Stöhr P (1934) Mikroskopische Studien zur Innervation des Magen-Darmkanals. Z Zellforsch Mikrosk Anat 21:243–278

Stöhr P (1952) Zusammenfassende Ergebnisse über die mikroskopische Innervation des Magendarmkanals. Ergebn Anat Entwick 34:250–401

Sundler F, Håkanson R, Leander S (1980) Peptidergic nervous systems in the gut. Clin Gastroenterol 9:517–543

Sundler F, Håkanson R, Leander S, Uddman R (1982) Neuropeptides in the gut wall: cellular and subcellular localization, topographic distribution and possible physiological significance. In: Chan-Palay V, Palay SL (eds) Cytochemical methods in neuroanatomy. Alan R Liss, New York, pp 341–356

Sundler F, Moghimzadeh E, Håkanson R, Ekelund M, Emson P (1983) Nerve fibers in the gut and pancreas of the rat displaying neuropeptide-Y immunoreactivity. Intrinsic and extrinsic origin. Cell Tissue Res 230:487–493

Surprenant A (1984a) Slow excitatory synaptic potentials recorded from neurones of guinea-pig submucous plexus. J Physiol (Lond) 351:343–361

Surprenant A (1984b) Two types of neurones lacking synaptic input in the submucous plexus of guinea-pig small intestine. J Physiol (Lond) 351:363–378

Taguchi T, Tanaka K, Ikeda K, Matsubayash S, Yanaihara N (1983) Peptidergic innervation irregularities in Hirschsprung's disease. Virchow's Arch [A] 401:223–235

Tanaka T, Starke K (1979) Binding of ^{3}H-clonidine to an α-adrenoceptor in membranes of guinea-pig ileum. Naunyn-Schmiedeberg's Arch Pharmacol 309:207–215

Tange A (1983) Distribution of peptide-containing endocrine cells and neurons in the gastrointestinal tract of the dog: immunohistochemical studies using antisera to somatostatin, substance P, vasoactive intestinal polypeptide, met-enkephalin, and neurotensin. Biomed Res 4:9–24

Tapper EJ (1983) Local modulation of intestinal ion transport by enteric neurons. Am J Physiol 244:456–468

Tapper EJ, Lewand DL (1981) Actions of a nicotinic agonist, DMPP, on intestinal ion transport in vitro. Life Sci 28:155–162

Tapper EJ, Powell DW, Morris SM (1978) Cholinergic-adrenergic interactions on intestinal ion transport. Am J Physiol 235:E402–E409

Tapper EJ, Bloom AS, Lewand DL (1981) Endogenous norepinephrine release induced by tyramine modulates intestinal ion transport. Am J Physiol 241:G264–G269

Taub M, Bonorris G, Chung A, Coyne MJ, Schoenfield LJ (1977) Effect of propranolol on bile acid- and cholera enterotoxin-stimulated cAMP and secretion in rabbit intestine. Gastroenterology 72:101–105

Taylor IL, Vaillant CR (1983) Pancreatic polypeptide-like material in nerves and endocrine cells of the rat. Peptides 4:245–253

Temesrékási D (1955) Die Synaptologie der Dünndarmgeflechte. Acta Morphol Acad Sci Hung 5:53–69

Thomas EM, Templeton D (1981) Noradrenergic innervation of the villi of rat jejunum. J Auton Nerv Syst 3:25–29

Tidball CS (1961) Active chloride transport during intestinal secretion. Am J Physiol 200:309–312

Tidball CS, Tidball ME (1958) Changes in intestinal net absorption of a sodium chloride solution produced by atropine in normal and vagotomized dogs. Am J Physiol 193:25–28

Tien X-Y, Wahawisan R, Wallace LJ, Gaginella TS (1985) Intestinal epithelial cells and musculature contain different muscarinic binding sites. Life Sci 36:1949–1955

Trent DF, Weir GC (1981) Heterogeneity of somatostatin-like peptides in rat brain, pancreas, and gastrointestinal tract. Endocrinology 108:2033–2038

Tsai BS, Conway RG, Bauer RF (1985) Identification and regulation of alpha$_2$-adrenergic receptors in rabbit ileal mucosa. Biochem Pharmacol 34:3867–3873

Tsuto T, Iwai N, Yanagihara J, Majima S, Ibata Y (1983) Immunohistochemical investigation of vasoactive intestinal polypeptide (VIP) and substance P in the colon in Hirschsprung's disease. Jpn J Ped Surg 19:13–20

Turnberg LA, McKay J, Higgs J (1982) The role of opiates in the control of small intestinal transport. In: Case RM, Garner A, Turnberg LA, Young JA (eds) Electrolyte and water transport across gastrointestinal epithelia. Raven Press, New York, pp 287–294

Udall JN, Singer DB, Huang CTL, Nichols BL, Ferry GD (1976) Watery diarrhea and hypokalemia associated with plasma vasoactive intestinal peptide in a child. J Pediat 89:819–821

Ussing H, Zerahn K (1951) Active transport of sodium as the source of electric current in the short-circuited isolated frog skin. Acta Physiol Scand 23:110–127

Valiulis E, Long JF (1973) Effects of drugs on intestinal water secretion following cholera toxin in guinea pigs and rabbits. Physiologist 16:475

Vinayek R, Brown DR, Miller RJ (1983) Inhibition of the antisecretory effects of [D-Ala^2, D-Leu^5]enkephalin in the guinea-pig ileum by a selective delta opioid antagonist. Eur J Pharmacol 94:159–161

Vinayek R, Brown DR, Miller RJ (1985) Tolerance and cross-tolerance to the antisecretory effects of enkephalins on the guinea-pig ileal mucosa. J Pharmacol Exp Ther 232:781–785

Vincent SR, Dalsgaard C-J, Schultzberg M, Hökfelt T, Christenson I, Terenius L (1984) Dynorphin-immunoreactive neurons in the autonomic nervous system. Neuroscience 11:973–987

Vinik AI, Gaginella TS, O'Dorisio TM, Shapiro B, Wagner L (1981) The distribution and characterization of somatostatin-like immunoreactivity in epithelial cells, submucosa, and muscle of the rat stomach and intestine. Endocrinology 109:1921–1926

Waddell MC (1929a) A histological study of the enteric plexuses in the small intestine following degeneration of the extrinsic nerves. Anat Rec 42:65

Waddell MC (1929b) Anatomical evidence for existence of enteric reflex arcs following degeneration of extrinsic nerves. Proc Soc Exp Biol Med 26:867–869

Wade PR, Westfall JA (1985) Ultrastructure of enterochromaffin cells and associated neural and vascular elements in the mouse duodenum. Cell Tissue Res 241:557–563

Waldman DB, Gardner JD, Zfass AM, Makhlouf GM (1977) Effects of vasoactive intestinal peptide, secretin and related peptides on rat colonic transport and adenylate cyclase activity. Gastroenterology 73:518–523

Walling MW, Brasitus TA, Kimberg DV (1977) Effects of calcitonin and substance P on the transport of Ca, Na, and Cl across rat ileum in vitro. Gastroenterology 73:89–94

Walter P (1956) Das morphologische Verhalten vegetativ-nervöser Elemente im Duodenum des Rindes. Acta Neurol 15:79–100

Warhurst G, Smith G, Tonge A, Turnberg L (1983) Effects of morphine on net water absorption, mucosal adenylate cyclase activity and PGE$_2$ metabolism in rat intestine. Eur J Pharmacol 86:77–82

Warhurst G, Smith GS, Higgs N, Tonge A, Turnberg LA (1984) Influence of morphine tolerance and withdrawal on intestinal salt and water transport in the rat in vivo and in vitro. Gastroenterology 87:1035–1041

Wright RD, Florey HW, Jennings MA (1938) The secretion of the colon of the cat. Q J Exp Physiol 28:207–229

Wright RD, Jennings MA, Florey HW, Lium R (1940) The influence of nerves and drugs on secretion by the small intestine and an investigation of the enzymes in intestinal juice. Q J Exp Physiol 30:73–120

Wu ZC, O'Dorisio TM, Cateland S, Mekhjian HS, Gaginella TS (1979) Effects of pancreatic polypeptide and vasoactive intestinal peptide on rat ileal and colonic water and electrolyte transport in vivo. Dig Dis Sci 24:625–630

Yada T, Okada Y (1984) Electrical activity of an intestinal epithelial cell line: hyperpolarizing responses to intestinal secretagogues. J Membr Biol 77:33–44

Yamaguchi K, Abe K, Miyakawa S, Ohnami S, Sakagami M, Yanaihara N (1980) The presence of macromolecular vasoactive intestinal polypeptide (VIP) in VIP-producing tumours. Gastroenterology 79:687–694

Yanaihara C, Sakagami M, Michuzuki T, Sato H, Yanaihara N, Iwanaga T, Fujii S, Fujita T (1980) Immunoreactive VIP (vasoactive intestinal polypeptide) in canine intestinal mucosa and muscle. Biomed Res 1:449–455

Yanaihara N, Nokihara K, Yanaihara C, Iwanaga T, Fujita T (1983) Immunocytochemical demonstration of PHI and its coexistence with VIP in intestinal nerves of the rat and pig. Arch Histol Jpn 46:575–581

Zimmerman TW, Binder HJ (1982) Muscarinic receptors on rat isolated colonic epithelial cells: a correlation between inhibition of (^{3}H)-quinuclidinyl benzilate binding and alteration in ion transport. Gastroenterology 83:1244–1251

Zimmerman TW, Binder HJ (1983) Effect of tetrodotoxin on cholinergic agonist-mediated colonic electrolyte transport. Am J Physiol 244:G386–G391

Zimmerman TW, Dobbins JW, Binder HJ (1982) Mechanism of cholinergic regulation of electrolyte transport in rat colon in vitro. Am J Physiol 242:G116–G123

Rev. Physiol. Biochem. Pharmacol., Vol. 109
© by Springer-Verlag 1987

Exercise Training and Its Effect on the Heart

DAVID A. S. G. MARY

Contents

Department of Cardiovascular Studies, The University, Leeds LS2 9JT, United Kindom

1 Introduction

There have been many investigations involving the effects on the heart of physical training by repeated dynamic muscular exercise. In several reviews, interest has been shared by basic scientists, health experts and clinicians and has largely focused on the possibility of beneficial effects, particularly in terms of the prevention or treatment of ischaemic heart disease (e.g. Ekelund 1969; Froelicher 1972; Rowell 1974; Clausen 1976, 1977; Leon and Blackburn 1977; Scheuer and Tipton 1977; Greenberg et al. 1979; Stone 1980a; Wyatt 1982; Rigotti et al. 1983; Blomqvist and Saltin 1983; Froelicher 1983; Schaible and Scheuer 1985). In general, studies in animals have indicated the occurrence of training-related improvements in cardiac performance. In respect of coronary blood supply and myocardial ischaemia, however, the findings regarding the benefit of such improvements have not been consistent.

The reasons for such inconsistency are not unequivocally known. However, several factors could be implicated, which include the accuracy of techniques used, the possibility of a small benefit, interference by concomitant training effects, and variability related to the organism, training programmes and species differences, as has been previously reported (e.g. Schaper et al. 1972; Scheuer and Tipton 1977; Schaible and Scheuer 1985; Wyatt 1982). Many of the interfering factors are less readily controlled in studies in man or in conscious animals than in experimental preparations, when appropriate methods may be used and laboratory conditions specifically defined. Meaningful deductions from laboratory findings would in addition require careful consideration of the influence of experimental techniques (e.g. Linden and Mary 1983).

This review considers training-induced effects on the heart reported in experimental preparations and points out recent methods considered reliable in assessing the effects in man. Emphasis will be placed on reports involving changes in coronary blood supply and myocardial ischaemia. Although discussion of mechanisms underlying effects of training is beyond the scope of this review, some examples will be mentioned.

2 Effects of Training in General

As alluded to in the Introduction, laboratory experiments in animals offer the potential of demonstrating a direct effect of exercise training on the heart. In the context of coronary blood supply, it would be possible to control or define other concomitant training-related changes, e.g. changes in haemodynamic variables, which, by their effects on cardiac performance, coronary vasculature or circulatory reflexes, could influence coronary blood flow.

One of the most established haemodynamic effects of training has been on heart rate. Less consistent effects have been reported on blood pressure, volume and flow and cardiac mass and size, perhaps because of inaccuracy of measurement or interplay between direct and reflex consequences of the haemodynamic changes. This section will outline some of these effects of training.

2.1 Heart Rate

Several studies have established that physical training results in decreases in heart rate (see Schaible and Scheuer 1985). This bradycardia has been demonstrated in experimental animals, including the rat, cat, dog, miniature swine and horse (e.g. Marsland 1968; Wyatt and Mitchell 1974; Williams and Potter 1976; Dowell et al. 1977; Scheuer and Tipton 1977; Stone 1977; Sanders et al. 1978; Gleeson et al. 1983) and has also been reported in man (e.g. Saltin et al. 1968; Clausen 1977).

The decrease in heart rate has been shown during submaximal levels of workload in exercise tests (e.g. Williams and Potter 1976; Stone 1977; Sanders et al. 1978; Gleeson et al. 1983) but has not been consistently found at maximal levels of exercise (e.g. Saltin et al. 1968; Barnard et al. 1980). The occurrence of training-induced resting bradycardia has not been uniform: Though it has been possible to show it in some studies in experimental animals (e.g. Wyatt and Mitchell 1974; Sanders et al. 1978; Schaible and Scheuer 1979; Musch et al. 1985), it has not been demonstrated in others (e.g. Williams and Potter 1976; Bove et al. 1979; Stone 1980a) and has been considered too variable for accurate assessment of training (e.g. Tipton et al. 1974). It is difficult to rule out the possibility of effects related to the nervous system and consciousness of the experimental animals which might mask training-induced resting bradycardia. In one study, it was possible to show this effect only when the animal was asleep (Breisch et al. 1986). Anaesthesia is expected to impose its own interfering variables (e.g. Linden and Mary 1983), which could mask such a (probably small) training effect. There have been reports in which there were no significant differences in heart rate under anaesthesia between trained and sedentary animals, despite the presence of training bradycardia beforehand (e.g. Cohen et al. 1978; Sanders et al. 1978; Carey et al. 1983). It should be pointed out at this stage that the above reports have involved different animals with various ages and different training programmes. These findings do not refute the claim that training causes a decrease in heart rate; however, they suggest that this effect could differ in extent relative to the experimental design used to disclose it, as will be very briefly outlined below.

The reduction in heart rate has been mainly reported to occur during the initial 4−8 weeks of training by running, as demonstrated e.g. in rats, dogs and man (Tipton 1965; Wyatt and Mitchell 1974; Saltin et al. 1968; Siegel et al. 1970; Li et al. 1986), and to disappear 2–5 weeks after the cessation of training (Roskamm 1967; Wyatt and Mitchell 1974; Tipton et al. 1974). In studies of male and female rats trained by running on a treadmill or swimming, the decrease in the resting heart rate was found to be least consistent in the running female rats (Schaible and Scheuer 1981); this difference was one of several and was attributed to the sex of rats and mode of training. Though assessment using resting heart rate is subject to interfering variables, as mentioned above, it is interesting to point out some reported differences between running and swimming.

Swimming has been proposed to involve less energy expenditure (e.g. McArdle and Montoye 1966; McArdle 1967), but to impose more stress than running in the rat. Changes in environment and temperature and, possibly, respiratory problems and forced learning to swim have been thought to have their own effects, in addition to that of training (e.g. Baker and Horvath 1964; McArdle and Montoye 1966; McArdle 1967; Thomas and Millar 1958; Crews and Aldinger 1967). A decrease or increase in food intake has been reported in male or female rats respectively (Oscai et al. 1971 a, Harpur 1980), and thus a decrease in body weight has been more consistent in male than in female rats (see Schaible and Scheuer 1981). Any effect of forced training would be expected to manifest itself early in the training process (O'Brien 1981) and might be minimised by an adequate training period.

2.2 Heart Enlargement

Cardiac hypertrophy has been considered one of the effects of training. The occurrence of hypertrophy has been identified in trained as against sedentary groups of animals such as the rat, dog and pig by greater values for the following variables: cardiac or ventricular weight (e.g. Crews and Aldinger 1967; Barnard et al. 1980; Anversa et al. 1982; Breisch et al. 1986), these weights relative to body weight (e.g. Grimm et al. 1963; Ljungqvist and Unge 1972), ventricular wall thickness (e.g. Barnard et al. 1980; Anversa et al. 1983) and cross-sectional area of myocytes (Breisch et al. 1986). In longitudinal studies in intact animals or man, training-induced hypertrophy has been inferred from radiographic or electrocardiographic assessment of ventricular wall thickness changes (e.g. Wyatt and Mitchell 1974; Ehsani et al. 1978; de Maria et al. 1978). Such an inference should be considered together with expectations of the influence of changes in cardiac dimensions, wall contents of connective tissue and fluids. These aspects have been considered in experimental animals; the occurrence of excess fluid has been ruled out by assessing dry heart or ventricular weight (e.g. Leon and Bloor 1968; Schaible and Scheuer 1981). Furthermore, trained young male rats were reported to have greater cardiac proportions of myocardial fibres and sarcoplasmic mass, but similar interstitial space in comparison with sedentary rats (Bloor et al. 1970). Also, there have been indications that increases in connective tissue might not be of significance in repeated exercise training (Hickson et al. 1983). Assessment of ventricular volume is difficult, as will be mentioned in subsequent sections of this review; it is dependent on changes in haemo-dynamic variables.

The occurrence of hypertrophy has been reported to depend on the sex, age and species of animals and the mode of training. In the rat, hypertrophy has been consistently reported in female rats trained by swimming (e.g. Oscai et al. 1971b; Schaible and Scheuer 1981). Hypertrophy has been observed in young male rats trained by swimming 1 h twice weekly; adult rats required daily swimming, and old rats were considered to have lost myocardial fibres (Leon and Bloor 1968; Bloor and Leon 1970; Bloor et al. 1970). Hypertrophy is reported to occur in female rats within 2 weeks of swimming at a frequency of 75 min twice daily, 5 days/week (Buttrick et al. 1985a) or even within a shorter period (Hickson et al. 1979) and to progress with training until about the third week, when it alters only slightly (Buttrick et al. 1985a). Upon cessation of training in the rat, hypertrophy regresses in 3−7 weeks (Leon and Bloor 1968; Hickson et al. 1979, 1983; Buttrick et al. 1985a).

2.3 Other Effects

Other heart-related alterations in the body which have been examined in connection with training include changes in blood volume, arterial blood pressure, oxygen consumption, stroke volume and cardiac output. For instance, increases in plasma and red cell volumes during training have been reported in the dog (e.g. Musch et al. 1985) and in man (see Scheuer and Tipton 1977; Conventino et al. 1983; Coyle et al. 1986), though the increases are believed to be influenced both by the baseline value before training and by associated thermal stresses (see Harrison 1985). In longitudinal studies in animals and in man, training-related increases in cardiac output and oxygen consumption during maximal exercise have been reported, and decreases or variable changes occurred during submaximal exercise; the changes in resting conditions have not been systematic (e.g. Saltin et al. 1968; Wyatt and Mitchell 1974; Clausen 1976, 1977; Barnard et al. 1980; Gleeson et al. 1983; Mazzeo et al. 1984; Breisch et al. 1986). As would be expected from the reported training effects on heart rate and cardiac output, increases in stroke volume have been found before and during exercise. The effect of training on arterial blood pressure has been less certain (e.g. Clausen 1976, 1977; Scheuer and Tipton 1977).

The training-related changes, mentioned above, interact to a certain degree with each other, with other general training effects such as bradycardia and with changes in cardiac performance and coronary blood supply. These issues will be mentioned in some detail in subsequent sections of this review. In addition, peripheral changes are thought to occur, though their precise quantitative relationship to the heart has not been unequivocally established. For instance, changes have been reported in skeletal muscle enzymes, function and metabolic consequences (e.g. Holloszy and Coyle 1984), and reports involving animals and man have suggested a role in training effects for the nature of trained muscles or related indices such as oxygen consumption and lactate production (e.g. Baldwin et al. 1977; Clausen 1976, 1977; Musch et al. 1985; Parsons et al. 1985).

The above findings on training effects in general do highlight an important aspect: The accuracy and ease of measuring the heart rate and the certainty of the training-induced bradycardic effect place it in a prominent position amongst other training effects for use as an indicator to evaluate the effectiveness of the training intervention, particularly in longitudinal studies of changes related to the heart. Other, relatively less readily available indices of cardiac-related training effectiveness will be outlined later in this review.

3 Experimental Evidence

This section considers effects of training on the performance of the heart and coronary blood supply. It is expected that the general effects of training outlined in Sect. 2 might be involved in such considerations: for instance, that training might exert an indirect influence through its effect on variables such as the heart rate, ventricular dimensions, aortic blood pressure and flow (e.g. Leonard and Hajdu 1962; Folkow and Neil 1971; Feigl 1983) and their related reflex effects (e.g. Daly and Scott 1962; Kirchheim 1976; Vatner and Murray 1982; Feigl 1983). Furthermore, an interplay is expected between cardiac performance and coronary blood supply (e.g. Feigl 1983). The changes in cardiac performance and the possible interfering variables will initially be outlined.

3.1 Cardiac Performance

The reports on cardiac performance have involved changes in the inotropic state and the effect of the initial length of myocardial fibres, subject to additional influence related to haemodynamic variables; i.e. the Starling mechanism. These changes have been sought in myocardial tissue preparations, isolated perfused hearts and hearts of anaesthetised and conscious animals.

At this stage, a brief outline is warranted of indices used in the reported studies to be reviewed in this section.

In myocardial preparations such as the papillary muscle, various assessments have been used. In preparations made to contract at the same frequency of stimulation, stretching to increase preload length or tension results in an increase in isometrically developed tension. An increase in tension or the maximal rate of increase in tension (dT/dt max) or in force (dF/dt max) at the same initial length or lengths which lead to maximal resting tension (Lmax), may be obtained for example by means of the inotropic effect of catecholamines. In isotonically contracting muscles, the inotropic effects have been evaluated using the maximal velocity of shortening (Vmax), as derived from extrapolation of data to those at zero load and obtained with similar mechanical properties of the muscle. These indices have been extended for use in the heart. The effect of the initial length, or operation of the Starling mechanism, has been represented in terms of increases in ventricular stroke volume or external work during increases in ventricular filling pressure or volume. The inotropic effects have been assessed using various indices which include for example the maximal rate of development of ventricular pressure (dP/dt max), velocity of contractile element shortening (Vce) or velocity of circumferential fibre shortening (Vcf). The use of these indices in the assessment of cardiac performance has been previously reviewed in detail (e.g. Abbott and Mommaerts 1959; Sarnoff et al. 1960; Sonnenblick 1962; Leonard and Hajdu 1962; Sonnenblick et al. 1969; Folkow and Neil 1971).

3.1.1 Papillary Muscle

Preparations of papillary muscles or trabeculae carneae have been used to assess the effect of training on the intrinsic performance and physical properties of the myocardium (e.g. Nutter and Fuller 1977; Scheuer and Tipton 1977; Stone 1980a). Such preparations allow definition and control of variables, e.g. loading conditions, and unlike the case of the whole heart avoid interference by factors such as haemodynamic changes, geometrical assumptions and reflex effects. However, the preparations might involve the release of neurotransmitters and variability related to trauma (Allen 1983).

The reported effect of training on the contractile behaviour of the papillary muscle has not been consistent. It should be pointed out that the reported studies have involved comparisons between trained and sedentary groups of animals. The baseline values of variables before training in the same animals were unknown. Indices of contractile behaviour are more sensitive to changes in the same muscles than to differences between animals and are subject to variablility related to the influence of differences in the techniques used. Conclusions have depended on statistical analysis between groups of animals, such that adequate numbers would be required to reduce variability.

In three reports, male and female rats were trained by running on a treadmill up to 1 h daily or by swimming up to 6 h daily, 5 days/week for about 6–11 weeks (Tibbits et al. 1978, 1981; Mole 1978). Trained rats had greater gastrocnemius muscle cytochrome c oxidase activity (Tibbits et al. 1978, 1981) or ventricular dry weight (Mole 1978) than the sedentary control groups. In these reports, the left ventricular papillary muscle was examined. In trained male and female rats, there was no evidence of cardiac hypertrophy. The muscles were subjected to isometric twitch studies (Tibbits et al. 1978, 1981); peak developed tension or force and its dF/dt max at various bath calcium concentrations were greater in trained rats, but the response to increasing stimulation frequency was not significantly different. No significant differences were found between trained and sedentary groups in the time required to attain peak developed tension or the resting tension (Tibbits et al. 1978).

In the report of Mole (1978), in which swimming female rats showed evidence of ventricular hypertrophy, no differences were found in the relation of length to passive tension in muscle or in the time required to attain peak developed tension. In these isometrically contracting muscles, the developed tension during various preload values at and below Lmax and dT/dt max were greater in trained rats; the same applied to isoprenaline-induced increases in dT/dt max. In preparations for force-velocity analysis, the shortening velocity of muscle in lengths per second relative to resting values was greater at various loads in trained rats, as was Vmax at the lowest load. In this report, it was proposed that an adequate number of tests were required to minimise variability, which could mask a small improvement (Mole 1978).

These three reports (Tibbits et al. 1978, 1981; Mole 1978) make it seem probable that training resulted in an inotropic effect and an improvement in contractile behaviour during both isometric contraction and shortening, and the improvement was greater during isoprenaline stress. No evidence could be found of any deterioration in muscle performance during increases in preload length or tension. The improvement was present in the absence of significant differences in passive stiffness and time of tension development.

There have been other reports of studies in the rat which did not show an improvement. In one report (Nutter et al. 1981), young and adult male rats trained by running on a treadmill up to 1 h daily 5 times per week for about 12 weeks and then detrained for a further 6 weeks were compared with sedentary control rats. Trained rats had greater gastrocnemius muscle succinate dehydrogenase activity, but no evidence of cardiac hypertrophy. In isometrically contracting left ventricular papillary muscles, no statistically significant difference was observed in length-passive tension curves or time of tension development. The peak developed tension and its dT/dt max were lower in trained young rats and not significantly different in adult rats; also, no differences were found in calcium- or noradrenaline-induced increases in peak developed tension. It is remarkable that the young rats showed training-related deteriorations in contractile behaviour which did not occur in adult rats and that this deterioration was reversed by detraining relative to the sedentary rats. As will be mentioned in subsequent sections of this review in the rat, the period of 6 weeks is similar to that during which any detraining-related reversal of cardiac improvements other than in coronary structure would occur (e.g. Tepperman and Pearlman 1961; Leon and Bloor 1968; Hickson et al. 1979, 1983). However, no force-velocity analysis was made (Nutter et al. 1981), and the intensity of running on the treadmill during training was less than that in the reports showing improvement (Mole 1978; Tibbits et al. 1981); indeed, positive evidence of cardiac training effects was lacking (Nutter et al. 1981), though the possibility of a mild hypertrophy cannot be excluded in such cross-sectional studies.

In another report, male rats with or without aortic constriction were trained by running on a treadmill up to 4 h daily 5 days/week for about 8 weeks and were compared with sedentary groups to examine the influence of ventricular hypertrophy (Grimm et al. 1963). No evidence of ventricular hypertrophy was present in trained rats without aortic constriction, though their ventricular papillary muscles were heavier. No differences were found in the relations of length to passive or to developed tension in isometrically contracting muscles. In rats with aortic constriction and ventricular hypertrophy, the maximal developed tension was less than that in young sedentary rats without constriction, and training did not further reduce developed tension. The effects on rate of tension development were not studied, and there was no force-velocity analysis. Furthermore, as in the report of Nutter et al. (1981), training appeared less intensive than that reported by Mole (1978) and Tibbits et al. (1981). Of note in this report (Grimm et al. 1963) was the absence of deterioration in preload-related tension development during training.

The question of the effect of training on the performance of cardiac muscles has been reviewed (e.g. Nutter and Fuller 1977; Stone 1980a). Two further studies were cited involving isometric preparations of left ventricular papillary muscles or trabeculae obtained from rats with swimming-induced cardiac hypertrophy. No significant change in contractile performance of the papillary muscles and either improvement or no change in the trabeculae were said to have been observed in comparison with sedentary rats (Nutter and Fuller 1977; Stone 1980a). In the study of papillary muscles, the decrease in peak isometric tension noted during hypoxic conditions was less in trained than untrained rats (Amsterdam et al. 1973).

The above findings include data which suggest an influence of the age of rats; a brief consideration of this aspect is warranted, since various training-related changes in biochemical, cellular, vascular and cardiac performance indices have been related to the age of rats studied (e.g. Bloor and Leon 1970;

Scheuer and Tipton 1977; Capasso et al. 1982; Starnes et al. 1983; Mazzeo et al. 1984; MacIntosh et al. 1985), and age-related alterations in cellular biochemistry and performance of the heart are believed to occur (e.g. Hansford 1978; Templeton et al. 1979; Lakatta and Yin 1982; Capasso et al. 1982). It has been argued that age-related shifts in baseline values could influence the magnitude of training-induced haemodynamic changes (Starnes et al. 1983; Mazzeo et al. 1984).

In respect of the effects of age on myocardial performance, one report involved studies of left ventricular papillary muscles from female rats (Capasso et al. 1982). With progress of age, no changes were found in isometric preparations at resting tension or dT/dt max, but increases occurred in peak developed tension and the time to peak tension. During force-velocity analysis in isotonic preparations, the shortening velocity decreased. In another report, left ventricular trabeculae carneae were examined in adult and senescent male rats (Spurgeon et al. 1983). In isometrically contracting preparations, no differences were attributed to age in resting tension, peak developed tension and its dT/dt max; however, senescent rats had longer contraction times and greater dynamic stiffness.

The issue of interplay between age and training effects was also involved in the report of Spurgeon et al. (1983). The adult and senescent male rats were trained by running on motorised wheels 30 min daily 5 days/week for 18–22 weeks and were compared with sedentary control groups. Left ventricular hypertrophy in the trained groups was significant only in terms of ventricle/body weight ratio in adult rats. In trabeculae preparations, no effects were attributed to training in adult rats. Trained senescent rats did not differ from sedentary ones in resting tension, developed tension or its dT/dt max, but had lower times during contraction and dynamic stiffness coefficients which were similar to those in the younger rats. Dynamic stiffness was assessed by superimposing length changes in the preparation, and the coefficient represented the rate of change in stiffness relative to that in tension during contraction (Spurgeon et al. 1983).

In another report, similar groups of male rats were examined; training comprised running on a treadmill at intensities considered to be normalized for age, 5 days/week for about 20 weeks (Li et al. 1986). Assessments of heart rate and systolic arterial blood pressure before and during exercise were considered to have entailed resting bradycardia only in the trained young adult rats, and no differences in ventricular weights were reported. In isometrically contracting right ventricular papillary muscles, training did not significantly affect developed tension or its dT/dt max relative to developed tension in adult or aged rats; however, contraction times were greater in trained young adult rats and in sedentary aged rats than other groups of animals (Li et al. 1986).

These findings indicate that age-related effects on baseline values of intrinsic myocardial performance or physical properties are to be expected. There were indications that such baseline effects could have an influence on observed effects of training, either directly or concomitantly through alterations in physical myocardial properties. It is not possible unequivocally to explain the effect of age in the reports on training in the rat which were reviewed in this section, particularly those of Grimm et al. (1963) and Nutter et al. (1981), since such age aspects were not examined.

In the cat, as in the rat, some, but not all studies report having demonstrated exercise-induced improvements in papillary muscle preparations. In one report, cats were trained by swimming 45 min daily 5 days/week for about 20 weeks and compared with sedentary control cats; there were no significant differences in the heart weight (Wyatt et al. 1978). In preparations of isometrically contracting right ventricular papillary muscles, the means of cross-sectional areas and peak developed force at Lmax were greater in trained cats, and there were no significant differences in dF/dt max, time to peak force and increases in muscle performance during increases in length or frequency of stimulation. An improvement in contractile performance was also obtained during the addition of isoprenaline and was associated with a decrease in time to peak force. Normalisation for differences in cross-sectional area abolished the statistically significant difference in untreated muscles, though the trend of improvement remained (Wyatt et al. 1978). However, an improvement could be demonstrated under the influence of isoprenaline.

In another report, cats were trained by running on a treadmill to cause fatigue for 45 – 60 min 5 days/week for about 6 weeks; in contrast to a sedentary control group of cats, submaximal and maximal heart rates were lower, though heart weights were not different (Williams and Potter 1976). In isometrically contracting right ventricular papillary muscles, the increases in passive or developed force resulting from increases in length, time to peak force and dF/dt max were not significantly different, though mean developed tension at Lmax and its rate tended to be greater. In a further preparation, force-velocity analysis showed no differences between the two groups at muscle preloads which were not statistically significantly different (Williams and Potter 1976).

It could be concluded from the reviewed reports, which mainly involve the rat, that an improvement in intrinsic contractile performance is possibly related to training. Inotropic effects and improvement in contractility during shortening have been shown, and there was no evidence of any deterioration in the response of the myocardium to increases in preload length or tension. The improvements were reported in male and female rats with or without cardiac hypertrophy. Though the number of reports is not large, indications could be found that mild training might not have been effective. The extent of improvement in contractile performance was probably small, such that its demonstration was thought to be made possible either by imposing stressing conditions or by using adequate numbers of tests statistically to account for

interfering variables. Such variables have included differences in the techniques used in myocardial tissue preparations and the reported possibility of
structural changes attributed to bouts of forced exercise in some animals of
the study groups (Loguens and Gomez-Dumm 1967; Tomanek and Banister
1972). Uncertainty still remains regarding the possible influence of age
operating either directly through its effects on baseline contractile performance or concomitantly with training through the influence of age-related
changes in myocardial physical properties.

3.1.2 Heart

To examine the effect of training on the whole heart in experimental animals,
different methods have been used, which can conveniently be grouped in this
review according to whether they involve isolated hearts, anaesthetised
preparations or instrumented conscious animals. In general, the use of the
whole heart could be considered, in contrast to isolated parts of it, to allow
assessment of cardiac performance in terms of well-recognised and commonly used haemodynamic variables. In this context, it is important to outline
some fundamental differences between the three methods of assessing the effects of training.

Preparations of the heart in isolation from the body and circulation allow
direct assessment and definition of variables such as cardiac dimensions, rigid
control of interfering variables such as cardiac frequency and loading components and exclusion of any concomitant influence of the multitude of reflex
neural and hormonal effects and their arcane interactions. The heart in
anaesthetised animals could be argued to provide net effects of most of the
above aspects and include the influence of anaesthetic agents. In conscious
animals, the net effects would include those from higher centres of the nervous system, but longitudinal assessments and knowledge of pretraining levels
of some variables are more readily available, so some drawbacks of cross-sectional studies as outlined in the preceding section are thus avoided.

Isolated Perfused Heart

Cardiac performance has been assessed in isolated hearts which were perfused
with a modified Krebs-Henseleit solution (Penpargkul and Scheuer 1970).
Studies have been reported with regard to the effect of exercise training on rat
heart (e.g. Scheuer and Tipton 1977; Schaible and Scheuer 1985); of these, the
more recent studies have allowed assessments of end-diastolic volume and
aortic flow (Bersohn and Scheuer 1977). Throughout experiments, hearts
were electrically paced at a constant rate of 340 beats per minute and aortic

pressure kept constant; studies were made possible with preset levels of left atrial pressure.

Hearts from male rats trained by swimming or by running on a treadmill 75 min twice daily 5 days/week for about 8 weeks were compared with those from groups of sedentary control rats (Penpargkul and Scheuer 1970; Scheuer et al. 1974; Bersohn and Scheuer 1977; Giusti et al. 1978; Schaible and Scheuer 1979). Trained rats had lower heart rates at rest and during exercise (Scheuer et al. 1974; Schaible and Scheuer 1979) but in general displayed no evidence of cardiac hypertrophy; differences in fluid contents of myocardial wall were considered to be small (Bersohn and Scheuer 1977). Briefly, trained rats mainly had greater cardiac output, stroke volume, ejection fraction, stroke work, maximal power, extent of circumferential fibre shortening and Vcf at the same left atrial pressures; peak left ventricular pressures and dP/dt max were also greater, and there were no systematic differences in end-diastolic pressures or volumes. In the trained rats, there was a tendency for measured variables to show greater increases for the same increments in left atrial pressure. The isolated perfused heart was considered to provide more consistent training-induced changes than occurred in anaesthetised animals (Schaible and Scheuer 1979).

These findings indicated a training-related occurrence of greater left ventricular pump performance and contractility, particularly during ventricular ejection. The findings may be considered to extend to the heart improvements in papillary muscle performance reviewed in the preceding section, as reported by Tibbits et al. (1978, 1981) and Mole (1978); it is notable that the training programmes in the isolated heart series were in general shorter or less intensive than those in the three reports on papillary muscles.

With the isolated perfused heart, differences have been reported in training effects between male and female rats for which similar training programmes were employed (Schaible et al. 1981). Training comprised running on a treadmill up to 2 h per day 5 days/week for about 12 weeks; this running was therefore longer and more intensive than in the case of the male rats reviewed above. Sedentary control groups included rats subjected to food restriction to maintain body weights similar to those of male rats trained by running; this was considered necessary because variables normalised for left ventricular weight were greater in small than in large hearts (Schaible et al. 1981). Trained rats had greater gastrocnemius cytochrome oxidase activity; the differences were similar in male and female rats, and in neither group were there differences in dry heart weights. Trained male rats had greater cardiac output, stroke volume and stroke work at the same left atrial pressures and showed greater increases during increments in left atrial pressure. In contrast, no significant differences were found in female rats (Schaible et al. 1981).

These findings clearly show differences in training effects between male and female rats and carry important implications. In female rats, skeletal muscle indices of training effectiveness were not associated with a discernible cardiac training effect; this issue assumes relevance in respect of the contribution of peripheral training effects to those in the heart, as was alluded to in Sect. 2.3. However, despite indications of equivalent training duration and intensity and absence of cardiac hypertrophy in female rats, an improvement has been reported in the performance of papillary muscles (Tibbits et al. 1981), but not in the isolated perfused heart (Schaible et al. 1981).

The differences between male and female rats have been related to whether training involved swimming or running in female rats and to the occurrence of cardiac hypertrophy. Female rats were studied (Schaible and Scheuer 1981) in a swimming training programme similar to that used for male rats (e.g. Schaible and Scheuer 1979). Trained female rats had greater dry ventricular weights, which was not observed in trained male rats. In isolated perfused heart preparations from trained female rats, the cardiac output at similar left atrial pressures was greater than in preparations from sedentary female rats, though this difference was abolished when cardiac weight was taken into account. However, increases in cardiac output per increment in left atrial pressure were greater, over lower ranges of this pressure, in trained rats. Trained female rats showed greater stroke work, ejection fraction, peak left ventricular systolic pressure, extent of circumferential fibre shortening and peak Vcf at the same left atrial pressure; no difference was reported in dP/dt max. As a finding possibly related to cardiac hypertrophy in trained female rats, it was noted that no differences occurred relative to sedentary rats in left ventricular end-diastolic pressure or volume at all left atrial pressures at a time when volumes, normalised for ventricular weight, were smaller (Schaible and Scheuer 1981).

The finding that no improvement in cardiac performance occurred in female rats unless it was accompanied by ventricular hypertrophy contrasts with reports of improvements in papillary muscle from female rats trained by swimming or running in the presence or absence of cardiac hypertrophy (Mole 1978; Tibbits et al. 1981). However, differences were apparent in the length and intensity of the training programmes, as alluded to above.

There have been other reports which included examination of the influence of age on the training effects in the isolated perfused heart. In one report, aged male rats were trained by running on a treadmill up to 35 min daily 5 days/week for about 16 weeks and were compared with two sedentary control groups of male rats, young and aged (Starnes et al. 1983). No differences were found in the weight of the heart, and in all rats under a paced constant heart rate of 300 beats per minute, data were obtained before and after stresses induced by raising left atrial and aortic pressure. Peak systolic pressure, cardiac output and stroke volume were lower in aged than in young rats. In the group of aged rats, trained animals showed greater peak systolic pressure and cardiac output during the stress but were not different from young sedentary rats (Starnes et al. 1983).

In a further study using the isolated perfused heart (Fuller and Nutter 1981), young and old rats were trained by running as described in the report of Nutter et al. (1981) and mentioned in Sect. 3.1.1. Hearts were paced at a constant rate of 360 beats per minute and studied at various left atrial pressures. Trained rats had greater gastrocnemius muscle succinate dehydrogenase activity and, based on the heart/body weight ratio, were considered to have probable cardiac hypertrophy. With the hearts in the arrested state, however, left ventricular volume increments led to similar pressures in trained and untrained rats. Left ventricular pressure per volume was greater in old than in young rats. In the perfused heart examined at various left atrial pressures, no significant differences were found in cardiac output, left ventricular pressures, dP/dt max and stroke work; though statistically insignificant, trained rats showed trends of higher cardiac output and lower left ventricular pressures and dP/dt max in mean group data at high left atrial pressures. No differences were reported which were attributable to age (Fuller and Nutter 1981).

These findings do not refute the possibility of an age-related influence on the papillary muscles, as was construed in Sect. 3.1.1 nor do the findings rule out training-related improvements, as reviewed earlier in this section.

It could be concluded from this review of the isolated perfused heart that the prevailing findings indicated a possible training-related improvement in pump performance and contractility of the heart, particularly during ventricular ejection. In general, the findings were similar to those regarding the papillary muscle reviewed in the preceding section: Both involved a possible influence of age, a need to impose stresses to show changes and inconsistency in showing training-related improvements. Unlike the findings in papillary muscles, however, improvements in isolated hearts of female rats were related to the occurrence of cardiac hypertrophy and the design of the training programme. Taken together, it is possible that the findings indicated a small degree of improvement in relation to training.

Anaesthetised Animals

In one report, female rats trained by swimming up to 6 h daily 6 days/week for 118−250 h were compared with sedentary control rats (Crews and Aldinger 1967). Trained rats had greater heart weights and thicker ventricular walls. At similar arterial blood pressure and lower heart rate in the anaesthetised state, trained rats showed greater isometric systolic tension of the ventricular walls at initial tensions of 10−30 g, as assessed by a strain gauge lever system. The effects of intravenous adrenaline were reported to be weaker in trained rats, though they were associated with large changes in heart rate and blood pressure (Crews and Aldinger 1967). The findings might be considered to indicate preservation of the Starling mechanism, though further interpretations would take into account effects of the different haemodynamic variables.

In other reports in anaesthetised rats, cardiac stressing by increases in aortic blood pressure or alterations in the filling pressures of the heart were used to examine cardiac effects of training (Dowell et al. 1976; Codini et al. 1977; Cutilletta et al. 1979; Fuller and Nutter 1981). Male rats were trained by swimming 75 min twice daily 5 days/week for about 8 weeks (Codini et al. 1977) or by running on a treadmill 60 min daily 5 days/week for about 12 weeks and then detrained for 6 weeks (Fuller and Nutter 1981). Female rats were trained by running on a treadmill up to 60 min daily 5 days/week for 8 weeks (Dowell et al. 1976; Cutilletta et al. 1979). Relative to sedentary control groups, trained rats had greater gastrocnemius muscle succinate dehydrogenase activity and heart/body weight ratios (Fuller and Nutter 1981). Training-related improvement in cardiac performance was reported in three of these studies, mainly occurring during cardiac stress (Dowell et al. 1976; Codini et al. 1977; Cutilletta et al. 1979), but not in unstressed hearts (Fuller and Nutter 1981).

The improvement in trained male rats comprised greater peak left ventricular systolic pressures and dP/dt max at levels of late diastolic pressures increased by graded aortic constriction and at the same heart rate. Such improvement was still present when the same systolic pressures as in the control groups were selected for comparisons. In addition, the trained rats showed higher values for pressure and its dP/dt max during high levels of paced heart rates (Codini et al. 1977). In the other study of male rats, differences between groups of male rats were considered to have been produced by differences in haemodynamic variables (Fuller and Nutter 1981).

Trained female rats showed a greater cardiac contractility index, relating dP/dt max to pressure, following sustained aortic constriction, which also resulted in lower left ventricular end-diastolic pressures. During infusion or withdrawal of blood to change cardiac preload, there was a tendency towards greater increases in cardiac output and stroke volume or smaller decreases in these values respectively (Dowell et al. 1976; Cutilletta et al. 1979).

These findings support previous evidence in this review that cardiac stressing is required to show a possibly small improvement in cardiac pump performance and contractility. Furthermore, the possibility was raised that differences in haemodynamic variables could be argued to mask such a small improvement. Improvements were shown in female rats despite the shorter and less intensive running training programme used by Dowell et al. (1976) and Cutilletta et al. (1979). This is in contrast to the one employed by Schaible et al. (1981), which was not associated with an improvement in the isolated heart.

There have been other reports of studies involving various types of cardiac stressing, and some examples are outlined here. In one study in male rats, hypoxic ventilation or complete occlusion of the left coronary artery was used (Carey et al. 1976). Training involved running on a treadmill up to 90 min daily 5 days/week for 10–16 weeks. Relative to a weight-matched sedentary group, trained rats had higher gastrocnemius muscle cytochrome c oxidase activity. During hypoxia, trained rats were better able to maintain the heart rate, left ventricular systolic pressure and its dP/dt max. No differences were observed during coronary artery occlusion, which was attributed to the drastic nature of the occlusion (Carey et al. 1976). In another study on male rats, volume loading by dextran infusion with or without a more severe degree of hypoxic ventilation than in the report of Carey et al. (1976) was used (Yipintsoi et al. 1980). Trained rats were made to swim 75 min twice daily 5 days/week for 10 weeks; there were no differences in the weights of the heart between trained and sedentary rats. Comparisons between the two groups of rats were reported to show that the cardiac index of trained rats was better maintained, a finding which was attributed to their lower body weight and not to training (Yipintsoi et al. 1980). No significant benefit was found for trained rats; as in the report of Carey et al. (1976), from the data of Yipintsoi et al. (1980) the possibility could not be ruled out that a drastic intervention which depresses cardiac performance might mask a small improvement associated with training.

There are reports of stressing studies in dogs. For example, in one report, pressure or volume loading of the heart was used in beagle dogs (Bove et al. 1979). Training comprised running on a treadmill 75 min daily 5 days/week for about 8 weeks. Trained dogs had a slower heart rate during submaximal exercise tests than before training and an unchanged resting heart rate; these dogs also had greater heart/body weight ratios and gastrocnemius cytochrome c oxidase activity. At a constant heart rate controlled by atrial pacing, there were no significant differences in cardiac output or work during increases in arterial blood pressure by phenylephrine and during infusions of saline or dextran. The interventions did not significantly alter the cardiac output. Lack of sensitivity of the indicator-dilution technique, used to measure the cardiac output, in detecting small changes could have contributed to the findings (Bove et al. 1979), and the possibility of a mild degree of volume loading by infusion cannot be excluded.

It is relevant at this stage to highlight assessments of cardiac output in two subsequent reports (Ritzer et al. 1980; Carey et al. 1983), the details of which will be respectively reviewed in Sects. 3.1.2 and 3.2.2. In one report involving the effect of training on coronary resistance, the cardiac output measured by indicator-dilution technique at rest and during pacing-induced increases in heart rate was lower in a group of trained dogs than in a sedentary control group (Carey et al. 1983). In the other study, beagle dogs were examined by left ventricular angiography before and after training; at similar resting and paced heart rates, the stroke volume was found to increase, and the increase was statistically significant in the case of resting heart rate. In the same report, longitudinal comparisons in sedentary beagle dogs did not show a consistent change in stroke volume (Ritzer et al. 1980).

In another study, dogs were trained by running 20–25 min twice daily 4–5 days/week for about 8 weeks, after which their left ventricles weighed more than those of a sedentary control group (Riedhammer et al. 1976). The comparisons made between the two groups by means of angiography included left ventricular pressures, dP/dt max, Vce, volumes, stroke volume, ejection fraction and Vcf. They were repeated following vagotomy and administration of propranolol or during acute pressure loading by methoxamine at the same heart rate. Differences between the two groups of dogs were confined to the stressing test by acute pressure loading. Trained dogs had smaller increases in left ventricular end-diastolic pressure. Moreover, in contrast to sedentary dogs, there were no increases in end-diastolic volume and no deterioration in Vce at peak rate of pressure rise. An improvement in contractile performance during afterload stressing was construed on the basis of better performance from a smaller ventricular volume (Riedhammer et al. 1976).

The review in this section has shown inconsistent findings on training-related changes in performance of hearts in anaesthetised animals. Because of uncertainty regarding left ventricular dimensions, physical wall properties, reflex neural and hormonal effects and the cross-sectional nature of the studies, it is difficult to attribute a difference solely to training. However, some findings were consistent with those in the review on papillary muscles and isolated perfused heart. Together, they indicate a possible small improvement in cardiac performance, which might readily be masked by interfering haemodynamic variables and their reflex effects, by the absence of adequate stressing of the heart or by the use of drastic interventions which severely depress its performance.

Conscious Animals

Reports on the effects of training on the heart of conscious animals have mainly involved studies in the dog and include longitudinal analyses in animals before and after training, as well as cross-sectional comparisons between two groups of animals.

Reports of longitudinal studies in the dog have involved cardiac stressing by exercise or by increasing left atrial pressure before and after training (e.g. Dowell et al. 1977; Stone 1977; Ritzer et al. 1980; Musch et al. 1985).

In the reports of Dowell et al. (1977) and Stone (1977), training comprised running on a treadmill up to 75 min daily 5 days/week for about 8–10 weeks. This training was reported to result in increases in skeletal muscle cytochrome c oxidase activity (Stone 1977) and no differences in left ventricular weight between trained dogs and a sedentary control group (Dowell et al. 1977).

Training was planned to cause a decrease in heart rate during exercise tests. Cardiac output was assessed by electromagnetic flowmeters placed around the ascending aorta, with the late diastolic signal considered to represent zero flow, and had a variability of 7% between calibrations before and after studies. Pressures were assessed in the left atrium and ventricle. With dogs standing on the treadmill after training, the heart rate was lower and the cardiac output and dP/dt max were greater than before training; no significant changes were found in left ventricular systolic and end-diastolic pressures. In submaximal exercise tests, observations during training included decreases in heart rate, increases in cardiac output at high levels of exercise and increases in dP/dt max during exercise which were maintained when related to heart rates; no significant changes were observed in left ventricular pressures. Increases occurred during exercise in stroke volume, though they did not attain statistical significance (Dowell et al. 1977; Stone 1977). In some dogs, infusion was used to examine the effects of raising left atrial pressure from about 3 to 30 mmHg. After training, the left atrial pressure at which heart rate or cardiac output reached a plateau was less than that before training. These findings were considered, taking into account possible changes in ventricular dimensions and reflex effects, to indicate training-related improvements in left ventricular contractility and pump performance (Dowell et al. 1977; Stone 1977).

In the report of Musch et al. (1985), foxhounds were trained by running at 80% of maximal heart rate 60 min daily 5 days/week for about 8 – 12 weeks; in previously reported studies using similar training in dogs, decreases in heart rate and increases in left ventricular wall thickness were found, but not changes in left ventricular end-diastolic volume (Wyatt and Mitchell 1974). During training, increases occurred in plasma and red cell volumes (Musch et al. 1985). Maximal exercise tests were performed at times before and after training. A decrease in heart rate was shown only during submaximal exercise and an increase in cardiac output only during maximal exercise; no changes were found in arterial blood pressure. Oxygen consumption, derived using the cardiac output as assessed by dye dilution, was found to increase only during maximal exercise. The stroke volume, also derived using cardiac output, was shown to increase throughout the exercise test. In a previous study in dogs, stroke volume was measured using dye dilution and roentgenography of cardiac markers; similar group data were reported from the two measurements (Ordway et al. 1984). These findings were considered to represent training-related increases in maximal oxygen consumption, which were mainly due to increases in stroke volume (Musch et al. 1985).

The fourth report involved studies in beagle dogs, using submaximal exercise tests before and after training, which comprised running up to 75 min daily 5 days/week for 10 weeks; a sedentary group was also used for comparisons. Studies were performed at heart rates increased by electrical cardiac pacing and infusion and included assessments of left ventricular size (Ritzer et al. 1980). No evidence of cardiac hypertrophy was found, but skeletal muscle cytochrome c oxidase activity was greater in the trained dogs. There was a decrease in the heart rate during exercise, which mainly occurred within the first 5 weeks of training; no such decrease was seen during exercise in the sedentary group. Although the other assessments involved left ventricular angiography and pressure measurements made under anaesthesia, this study is reviewed in this section to put emphasis on its longitudinal nature. The inconsistent decrease in resting heart rate relative to that during exercise suggested central effects (Ritzer et al. 1980) and should recall similar interference, though involving different mechanisms, in assessments made in the anaesthetised state (e.g. Linden and Mary 1983). Significant changes during training whilst in sinus rhythm at an average heart rate of 103 beats per minute included increases in left ventricular end-diastolic volume, stroke volume and peak midwall stress in relation to pressure and volume. In the sedentary group, the only such change was a decrease in peak left ventricular systolic pressure. At paced heart rates averaging 190 beats per minute, significant increases during training were found in left ventricular peak systolic pressure, peak midwall stress, peak Vce and dP/dt max; no significant changes were found in the sedentary group. In only three dogs, during cardiac pacing and infusions to increase left ventricular end-diastolic volume, were the increases in stroke volume relative to those in end-diastolic volume similar before and after training. These findings were considered to suggest improvements in cardiac performance which were mainly obtained during the cardiac stress of increases in heart rate and training-related changes in ventricular dimensions. It was proposed, therefore, that comparisons involving trained and sedentary groups were not as sensitive as those involving longitudinal changes (Ritzer et al. 1980).

There have been other reports on conscious dogs in which trained groups of greyhounds were compared with untrained ones to assess the effect of training-related cardiac hypertrophy (Carew and Covell 1978) or changes in the coronary circulation (Restorff et al. 1977; Barnard et al. 1980). In the report of Carew and Covell (1978), ten conscious greyhounds considered to have been in a trained state were studied and compared mainly with 'normal' dogs from other studies. Trained animals had heavier hearts than two greyhounds which were less rigorously trained. In the trained animals, the heart rate was lower, and resting levels of left ventricular pressure, dP/dt max and Vcf did not differ from those obtained in normal dogs. During infusions to increase the left ventricular end-diastolic pressure, no significant differences between the groups were found in performance at a time when trained animals developed higher heart rates. The findings were considered to indicate preservation of "normal" performance of hearts with training-related hypertrophy (Carew and Covell 1978).

In the report of Barnard et al. (1980), exercise tests up to heart rates greater than 250 beats per minute in trained dogs were compared with those in sedentary dogs. Training involved running up to 2 h daily 5 days/week for 12–18 weeks. Trained dogs had greater left ventricular weights and gastrocnemius muscle maleate dehydrogenase activity. Trained dogs were found to have lower heart rates at rest and during submaximal, but not maximal exercise. The cardiac output, assessed using electromagnetic flowmeters around the ascending aorta, was greater only during maximal exercise; the stroke volume was greater throughout the exercise test in the trained dogs. Left ventricular dP/dt max was greater in trained dogs during maximal exercise, and differences in left ventricular systolic or end-diastolic pressures were not statistically significant (Barnard et al. 1980). In the report of Restorff et al. (1977), to be detailed in Sect. 3.2.2, the cardiac output measured by dye dilution was lower during exercise in the group of trained dogs; training lasted only 2 weeks, and values at the highest exercise test workloads were not given (Restorff et al. 1977).

Other reports have involved conscious rats or pigs (Gleeson et al. 1983; Breisch et al. 1986). Female rats were trained by running up to 60 min daily 5 days/week for 14–16 weeks. Running exercise tests were then performed, and the findings were compared with those in a sedentary group of rats (Gleeson et al. 1983). In the absence of differences in heart weight, trained rats had greater vastus lateralis muscle citrate synthase activity. The heart rate and cardiac output were lower in trained rats during both submaximal and maximal exercise. The oxygen consumption in trained rats was lower during submaximal and higher during maximal exercise (Gleeson et al. 1983). In the other report, maximal running exercise tests in pigs trained by running as described above for 12 weeks were compared with those in sedentary pigs (Breisch et al. 1986). Trained pigs had greater heart weights and larger myocytes. During maximal exercise, their heart rates and mean aortic blood pressure were similar to those in sedentary pigs, whilst the cardiac index and oxygen consumption were greater.

The above review indicates that it is possible to show training-related changes in conscious animals which are more consistent in longitudinal studies of the same animals and during cardiac stressing than in cross-sectional ones between trained and sedentary groups of animals and when such changes derived from unstressed hearts. Any training-induced improvements in cardiac pumping performance or contractility deduced through these studies were probably influenced by effects related to cardiac dimensions and reflex mechanisms. The review makes it possible to argue hypothetically that net improvements occurred during cardiac stressing conditions which were closer to the usual animal environment than those obtained in anaesthetised animals, isolated hearts or myocardial tissues. Furthermore, it is difficult to rule out the possibility that the improvements found in conscious animals, albeit possibly small, included those obtained in the isolated hearts or myocardium.

The above findings on training-related changes in haemodynamic variables, cardiac performance and dimensions and the associated reflex effects will assume relevance subsequently in Sect. 3.2.

3.2 Coronary Circulation

As in the case of the myocardium, the available reports concerning training-related changes in the coronary circulation have been inconsistent. The studies reported have included the function of the coronary circulation with both intact and narrowed or occluded vessels and consequences such as myocardial ischaemia or infarction. Training-induced changes have been sought in the structure of coronary vessels, blood flow, myocardial perfusion and their adequacy to meet the need of the myocardium for blood during stresses. The reported evidence will be reviewed according to these changes, though a variable interplay between them cannot be dismissed.

3.2.1 Structure of Coronary Vessels

This part of the review considers reports on the effect of training on coronary vascular structure in experimental animals. The reports have involved various parts of the coronary vasculature in animals in which the coronary vessels have not been narrowed, as well as the consequences of occluding the coronary arteries.

Intact Coronary Vessels

Essentially, studies reporting on the structure of the coronary vessels have involved anatomical and histological techniques designed to assess changes in the number of coronary vessels or their size. In general, the prevailing view has been that exercise training results in increases in the number of capillaries and the size of larger coronary vessels. This section will review training-related structural changes which have been reported in various segments of the coronary tree.

Myocardial Vessels. Several reports have involved comparisons between hearts from trained and sedentary control groups of animals regarding structure of vessels in the ventricular myocardium. Histological techniques were used mainly to examine vessels identified as capillaries with respect to differences in their number or proliferation. The number of capillaries has mainly been assessed in terms of capillary density in relation to myocardial sections or the ratio of capillaries to myocardial fibres; both assessments are believed to be influenced by training-induced ventricular hypertrophy (Hudlicka 1982).

In the rat, studies indicating that training results in increases in the density or the ratio of capillaries (e.g. Leon and Bloor 1968; Bloor and Leon 1970; Bloor et al. 1970; Tomanek 1970; Ljungqvist and Unge 1972; Bell and Rasmussen 1974; Leon and Bloor 1976; McElroy et al. 1978) have outnumbered those not reporting this finding (e.g. Parizkova et al. 1972; Anversa et al. 1982, 1983).

In one report in male rats, training involved swimming 1 h daily or twice weekly for 10 weeks, and the results were compared with a sedentary control group (Leon and Bloor 1968). Only rats trained daily had greater dry ventricular weights however, both trained groups, with or without ventricular hypertrophy, had greater capillary/fibre ratios, which persisted after 24–42 days of detraining (Leon and Bloor 1968). In another report, male rats were trained by swimming up to 1 h daily 5 days/week for 5 weeks (McElroy et al. 1978). In comparison with sedentary rats subjected to water immersion, there were no differences in ventricular weight or myocardial fibre diameter. The trained rats had greater capillary/fibre and capillary/ventricle ratios (McElroy et al. 1978).

The influence of age has been the subject of other reports. In two studies (Bloor and Leon 1970; Bloor et al. 1970), young, adult and old male rats were examined using methods similar to that reported by Leon and Bloor (1968). Greater dry ventricular weights were observed only in the young trained rats; the extent of this hypertrophy was greater in the group trained 6 days/week. In these studies, the greater capillary/fibre ratio in young rats was attributed to greater capillary numbers and in old rats to decreases in the number of myocardial fibres. In addition, in the young trained rats, a greater total number of capillaries, as well as the number per unit volume of myocardium, were found than in adult or old rats (Bloor and Leon 1970; Bloor et al. 1970). In another report, young, adult and old male rats were trained by running on a treadmill up to 40–50 min daily 6 days/week for 12 weeks and compared with sedentary groups (Tomanek 1970). Trained rats had decreases in resting and exercise heart rates, but there were no significant differences in myocardial fibre diameter. The three trained groups had a greater capillary density per unit area, which attained statistical significance only in the young rats. All trained rats had greater capillary/fibre ratios (Tomanek 1970).

The maintenance of training-induced changes in coronary structure in male rats has also been studied (Leon and Bloor 1976). In rats trained by swimming 1 h daily 5 days/week for 10 weeks the effects of decreasing the training regimen for a further 10 weeks were evaluated in relation to a group of rats kept without exercise for 20 weeks. With complete cessation of training, there were no differences relative to sedentary rats in indices of ventricular hypertrophy, which included dry ventricular weight, its proportion to body weight and sarcoplasmic contents, the number of myocardial fibres per unit area, the capillary density per unit volume and the capillary/fibre ratio. Decreases in swimming to 15 min five times weekly also abolished differences. Maintenance of ventricular hypertrophy required swimming 1 h five times weekly. In contrast, maintenance of capillary density required swimming 30 min twice weekly, and maintenance of capillary number and capillary/ fibre ratio required swimming 1 h five times weekly. Maintenance of a greater number of fibres required swimming either 30 min twice weekly or 1 h/week (Leon and Bloor 1976).

It has been suggested that a similar training-related change in coronary structure occurs, as compared with sedentary rats, in female rats trained by swimming 1 h daily 6 days/week for 3 months, attain a greater heart/body weight ratio. The capillary density was assessed visually in microscopic sections and was reported to be greater in the trained group (Ljungqvist and Unge 1972).

Capillary proliferation in relation to training in the rat has been assessed using an index related to the incorporation of labelled thymidine in endothelial nuclei. In one report, groups of female rats were trained by swimming 1 h daily 6 days/week for 2, 4 or 12 weeks or for 12 weeks followed by detraining for 2 weeks. Comparisons were then made with another group of normal control rats (Ljungqvist and Unge 1973). Heart/body weight ratios were reported to be greater in 2- and 4-week-trained than in control rats. Light-microscope autoradiography indicated neoformation of left ventricular capillary wall cells, relative to control rats, in rats trained for 2 weeks. This was further augmented at 4 weeks. No significant differences were found in the case of 12 weeks' training or after detraining; similar trends were reported in the right ventricle and in respect of connective tissue cells. The labelling index was also greater in left ventricular myocardial cells after 12 weeks' training (Ljungqvist and Unge 1973). These findings were complemented using electron microscopy in further, similar studies suggesting the training-related formation of capillaries. However, this was not consistently observed in female rats with cardiac hypertrophy caused by aortic stenosis (Mandache et al. 1972, 1973) or in female rats with aortic stenosis followed by a 4-week swimming period (Ljungqvist et al. 1976). In another study with a similar training programme, no capillary growth was observed in hypertrophied limb skeletal muscles (Ljungqvist and Unge 1977).

The same labelling technique has been used in male rats to examine the influence of age. Young, adult and old rats were trained by swimming 1 h daily 5 days/week for 3 weeks and compared with control groups (Unge et al. 1979); the trained rats had greater heart/body weight ratios. Capillary wall cell formation was reported to be greater only in young and adult rats (Unge et al. 1979).

In animals other than the rat, the reported effects of training on myocardial vessels have not been consistent. In one report, guinea pigs trained by running on a treadmill for about 15−30 min daily for 52 days were compared with sedentary ones (Tepperman and Pearlman 1961). Trained animals had no significant cardiac hypertrophy, though they developed anastomotic vessels between coronary arteries. In another report involving guinea pigs, strenuous training did not result in increases in capillary density (Hakkila 1955). In flying pigeons, evidence of cardiac hypertrophy and greater capillary density per unit area relative to nonflying littermates was reported (Rakusan et al. 1971).

In another report, young farm pigs were trained by running on a treadmill up to 60 min daily 5 days/week for 12 weeks; repeated exercise testing in these pigs showed an increase in maximal oxygen consumption and a decrease in sleeping heart rate. Moreover, in comparison with a sedentary group, the trained pigs showed evidence of left ventricular hypertrophy, in the form of greater weight and myocyte cross-sectional area (Breisch et al. 1986). Samples from the anterolateral free wall of the left ventricle of trained and sedentary pigs were examined; the number of capillaries per myocyte area in subendocardial and subepicardial layers was lower and the length of capillaries per myocyte volume was less in the trained pigs, though there were no differences in the diameter of the capillary lumen or in capillary endothelial area per myocyte volume. In contrast, the number of arterioles per myocyte area and their length per myocyte volume were greater in the trained than they were in the sedentary pigs (Breisch et al. 1986).

In one report with dogs trained by running on a treadmill 1 h daily at 39% – 72% of estimated maximal heart rate, 5 days/week for 12 weeks, biopsy samples from the right ventricular septal wall were examined. The number of blood vessels per unit area increased during training, a trend reversed by detraining for 5 weeks (Wyatt and Mitchell 1978). Also, training resulted in an increase in vessel perimeter when it was less than 13.4 μm and a decrease when it was greater than 15.5 μm beforehand (Wyatt and Mitchell 1978). It is notable in this longitudinal study that the knowledge of pretraining values permitted demonstration of a baseline effect.

These reports indicate that training was associated with increases in capillary density or ratio in male and female rats, whether trained by running or swimming and whether in the presence or absence of cardiac hypertrophy. The changes in capillaries are possibly associated with an increase in their number, which occurs more consistently with training in young than in old rats. A direct relationship was suggested between cardiac hypertrophy and frequency of training: increases in the number of capillaries start and continue at a lesser frequency or intensity of training than is the case with ventricular hypertrophy. Though fewer in number, reports on other animals have suggested variable results, including effects on vessels in the myocardium other than the capillaries.

Coronary Size. Studies in the rat have involved coronary segments which included vessels much larger than the capillaries. Injection of a polymer into the coronary vessels has been used to obtain a cast of the vessels occupied. The weight of a cast of the involved segment, which must be dependent at least on the injection pressure and the nature or content of the vascular segment, is then used to assess the size of a coronary tree. As in the case of coronary capillaries, reports on cast size involved comparisons between trained and matched sedentary rats.

In male and female rats trained by running daily for 5 weeks and male rats trained by swimming 30 min twice daily, 11 times over 6 days/week for 11 weeks, there was evidence of cardiac hypertrophy only in trained male rats (Tepperman and Pearlman 1961). The size of the casts was greater in the three trained groups than in control groups. The greater size of casts was also observed in groups of male rats trained by swimming as described above and then rested for 8 weeks, after which time cardiac hypertrophy was no longer present (Tepperman and Pearlman 1961).

In another report (Stevenson et al. 1964), groups of sedentary male rats were compared with groups trained by running on a treadmill 2 h daily, twice or five times weekly for 4 weeks or five times weekly for 2 weeks followed by 2 weeks' rest. Rats were also trained by swimming 1 h daily, twice or five times weekly for 4 weeks with or without rest and by swimming 1−4 h daily for 4 weeks. Evidence of hypertrophy was reported only in the strenuous 4-h swimming. Greater cast weights or cast/heart weight ratios were reported only in the case of moderate running which was not followed by rest and moderate swimming without rest (Stevenson et al. 1964). The reason for the lack of differences in cast weight, attributable to intensive training, is unknown. Similarly, in another study (Haslam and Stull 1974), greater cast weights observed in male rats following swimming twice weekly for 4 weeks were not augmented by swimming four times weekly or swimming at these intervals for 8 weeks.

Adolescent rats have been compared with adult ones; the rats were trained by running, with motivation using electical shock (Denenberg 1972). Greater cast weights or cast/heart weight ratios were not found in adolescent rats trained by running 1 h daily, 3 or 5 days/week for 5 weeks. Greater cast/heart weight ratios were observed in the only group of adult rats with a running frequency of 3 days/week. In none of the trained rats was cardiac hypertrophy reported (Denenberg 1972).

The findings in these reports suggest increases in the size of coronary tree in relation to training programmes, irrespective of the sex of rats or occurrence of cardiac hypertrophy. In this context, the improvement was similar to that observed in myocardial vessels (Sect. 3.2.1).

Large Coronary Vessels. Reports are available concerning the effect of training on the size of the lumen of large coronary arteries and of extracoronary collateral vessels, which connect systemic with coronary circulation (e.g.

Halpern and May 1958). These two vascular segments were examined in the reports of Leon and Bloor (1968, 1976), which were mentioned earlier in this Section (3.2.1). Comparisons of the sum of luminal cross-sectional area in the left and right coronary arteries 0.5 mm from their origin were made between trained and sedentary groups of young male rats. Swimming 1 h daily for 10 weeks was associated with a greater group average of luminal area, which, though not statistically significant, was of proportions similar to increases in the greater dry ventricular weights. Furthermore, a progressive decrease upon cessation of training occurred in parallel to that in ventricular weights. In contrast, neither ventricular hypertrophy nor greater luminal areas were found in young male rats made to swim twice weekly (Leon and Bloor 1968). In the same study, the luminal area of extracoronary collaterals was greater in the trained young male rats, regardless of the frequency of training or ventricular weight, and persisted along with the improved capillary/fibre ratio upon cessation of training and after regression of ventricular weight. In the report of Leon and Bloor (1976), maintenance of a greater extracoronary collateral area required swimming at least 1 h daily per week, in contrast to capillary density and ventricular hypertrophy, whose maintenance respectively required swimming 30 min daily, twice weekly and 1 h daily 5 days/week.

Regarding the age of rats, in the report of Bloor and Leon (1970), the extracoronary collateral area was found to show trends towards greater values in trained young, adult and old male rats. In contrast to this finding, the greater number or density of capillaries was found only in trained rats of a younger age.

Other reports are available in the dog (Wyatt and Mitchell 1978; Neill and Oxendine 1979). In the report of Wyatt and Mitchell (1978), dogs were trained by running on a treadmill 1 h daily at 39% – 72% of estimated maximal heart rate, 5 days/week for 12 weeks; this type of training has been suggested to lead to decreases in heart rate and increases in the thickness of left ventricular walls (Wyatt and Mitchell 1974). In longitudinal studies, contrast media were delivered to the coronary artery by aortic root injections. The diameter and cross-sectional area of proximal segments of the left circumflex artery, measured in angiograms at similar heart rates, were found to increase or decrease respectively during training and after detraining (Wyatt and Mitchell 1978).

The other report, which examined effects of training on occluded coronary vessels, as will be mentioned in Sect. 3.2.2, involved coronary angiography and is of interest to this section. Diameters of the proximal part of the left anterior descending artery and the posterior descending branch of the gradually occluded left circumflex artery were measured; also, visual assessments were made of collateral vessels between the two major arteries (Neill and Oxendine 1979). In comparison with a sedentary control group, dogs trained by running showed no evidence of cardiac hypertrophy, though the heart rate during exercise was lower after training. No significant differences were found in the diameter of the arteries or in the extent of collateral vessels; the latter result was attributed to a lack of sensitivity of the methods used (Neill and Oxendine 1979). The relevance of collateral blood flow will be considered in Sect. 3.2.2; at this stage, it is important to point out that proximal occlusion has been reported to result in a smaller collateral flow than distal occlusion (Schaper 1978; Reimer et al. 1981). Valid measurement of coronary calibre in plastic models could only be obtained in vessels greater than 3 mm in diameter (Bjork and O'Keefe 1976), which was greater than the average diameter encountered in the angiograms of Neill and Oxendine (1979).

The findings in the reports reviewed suggest that training results in changes in the coronary vasculature which differ according to the segment of the vascular tree. Any training-induced dilatation of large coronary arteries was related to ventricular hypertrophy, which was not found to correlate with improvement in the capillaries or medium-sized coronary tree, as mentioned in the preceding sections.

Coronary Occlusion

Other reports are available which involve assessment of training-related changes in the structure of cardiac infarction induced by coronary occlusion. Reports will be mentioned to provide examples of infarction inflicted before or following training.

In the absence of assessment of myocardial perfusion, it could be argued on the basis of structural observations that a link exists between myocardial vessels and progress of infarction. In one report, comparisons were made between cardiac infarction inflicted in trained and sedentary rats (McElroy et al. 1978). In this report, also mentioned earlier (Sect. 3.2.1), male rats trained by swimming had greater numbers of myocardial capillaries in the absence of cardiac hypertrophy. Myocardial infarction 2 days after occlusion of the left coronary artery was smaller in trained than in sedentary rats (McElroy et al. 1978).

In the case of training following infliction of myocardial infarction, structural assessment would at least include the influence of the severity of infarction, scar tissue and haemodynamic consequences. Examples will be considered in relation to the rat (Kloner and Kloner 1981; Musch et al. 1986) and the dog (Kalpinsky et al. 1968).

In the report of Kloner and Kloner (1981), the left coronary artery was occluded in rats, which were then divided according to electrocardiographic infarct size into training and sedentary groups. Training involved swimming up to 40 min daily for about 2 weeks. In the absence of differences in septal wall thickness, the healing scar tissue was less thick in the trained than in the sedentary rats (Kloner et al. 1981). Scar thinning in the healing phase was associated with training; however, it is not possible to distinguish any vascular effect from that of changes in scar or heart dimensions which could occur with such a drastic occulsion, as will be seen in the other report below.

The report of Musch et al. (1986) involved interventions in male rats consisting of ligation of the left coronary artery to cause myocardial infarction or sham intervention. After at least 6 weeks, each procedure was followed by assignment to sedentary groups or groups trained by running on a treadmill up to 1 h daily 5 days/week for 12 weeks. Trained rats had higher levels of succinate dehydrogenase activity in the soleus and plantaris muscles and achieved lower levels of lactate in the blood during exercise than sedentary rats. There were no differences in the weight of the heart, heart rate or maximal oxygen consumption during exercise. In this study, training was considered to be of moderate intensity, to have greater metabolic than cardiac effects and to have been imposed when fibrosis was possibly well established. Estimates of the size of infarction in trained rats did not differ from those in the sedentary group; i.e they were of a severity amounting to infarction of over a third of the left ventricle. In addition, myocardial infarction was thought to have caused cardiac failure, which was better tolerated by trained than sedentary rats. During maximal exercise testing, trained rats with infarction achieved higher heart rates and oxygen consumption relative to body weight than sedentary rats with infarction (Musch et al. 1986).

The report of Kalpinsky et al. (1968) involved the effect of training in conditions of established myocardial infarction in the dog. The left anterior descending artery was ligated, and after about 1 week, the dogs were assigned to a sedentary group or a group trained by running on a treadmill 30 min twice daily 6 days/week for about 4 weeks. Trained dogs had lower heart rates during rest and exercise on a treadmill, a lower cardiac index and smaller increases in blood lactate or plasma adrenaline levels during exercise. In the anaesthetised state, trained dogs showed a tendency towards lower left ventricular end-diastolic pressure. During selective left coronary angiography or in postmortem angiograms using gelatin barium sulphate, no differences in collateral vessels were shown. Histological studies of the myocardium established fibrous tissue formation and, whilst considered subject to shrinkage artefacts, did not show differences in the size of the infarction (Kalpinsky et al. 1968). These findings are essentially similar to those outlined above in the rat. In respect of the collateral vessels, the findings were reminiscent of those reported by Neill and Oxendine (1979), and their implications will be considered in Sect. 3.2.2.

The above review on coronary vascular structure, which mainly involved the rat, indicates that training possibly results in increases in the size of coronary vessels and number of capillaries.

In the case of intact coronary vessels, there were indications of an influence on the observed improvement exercised by age or possibly size of the heart and by the starting status of the vessels. Vascular inprovement appeared to be longer lasting and required less frequent training than the other effect of training, namely an increase in ventricular weight. There were differences in training effects between vascular segments of the coronary tree, which were even greater in animals other than the rat.

Regarding myocardial infarction induced by coronary occlusion, there were indications mainly in the rat that training was associated with reductions in the size of inflicted infarction. In established infarction, no training effect was shown on the size of scar tissue, though training-related benefits were found in terms of better cardiac haemodynamic performance, already depressed through cardiac infarction and failure.

Finally, it should be pointed out in general that the demonstration of training-related improvements in coronary vascular structure has been remarkable in view of expected interference by several factors. Variability is expected in histological techniques and their sensitivity, in pressures or flow of the vessels at the time of procurement and in the prevalent state of unstressed coronary circulation. Furthermore, in the case of myocardial infarction, interference is expected through the severity of infarction, scar tissue and haemodynamic consequences, including changes in dimensions of the heart.

The implications of the above considerations must, at least, include a degree of uncertainty in attributing structural improvements directly to training. Certainly, any quantification of increases in the number or size of coronary vessels would be subject to serious limitations.

3.2.2 Coronary Blood Flow

The review in the preceding section makes it relevant to consider the reported evidence on whether training results in improvements in coronary blood flow or myocardial perfusion in experimental animals with intact or narrowed coronary arteries. Clearly, any such relationship between changes in coronary structure and flow would involve several important issues. The techniques and design of studies used in examining the two changes are different. There could be an influence of concomitant changes in haemodynamic variables on structure and flow. In the case of narrowed or occluded coronary vessels, questions would arise involving the functional importance of coronary collateral vessels and extracoronary vascular connections and the degree of flow deprivation in terms of myocardial regions, layers or cells.

The above considerations will be included within the reported effects on coronary blood flow to be reviewed in this section; the reports will be grouped into two main parts concerning unoccluded and occluded coronary vessels.

Intact Coronary Vessels

In experimental animals in which the coronary vessels have not been narrowed, the available reports have not been consistent regarding training-induced improvements in coronary blood flow or myocardial perfusion. Some general considerations are required at this stage on the nature of reported studies.

Different techniques of blood flow assessment have been used, e.g. collection, flowmeters and labelled microspheres, the sensitivity of which has not always been defined. It could be argued that flow obtained by collection may be influenced by the method of assessment. In the case of flowmeters placed around or inside coronary vessels, limitations are expected, e.g. in terms of extending velocity measurements to flow, stability of zero flow recording, placement and contact of the probe relative to vessel walls and their effect on phasic pressure or flow waves (e.g. Kramer et al. 1963; Berne and Rubio 1979). Measurements using microspheres depend, for example, on their number in the region involved, trapping in relation to size and trauma, resolution of detecting techniques in relation to regional morphology, number of injections and background activity (e.g. Hoffman et al. 1983; Winkler 1984).

Perhaps of greater pertinence to this review is the difficulty of specifically relating findings on flow to training, since various cardiac and circulatory consequences of training are known to have variable effects on coronary blood flow which have not always been assessed. The extent of interplay between the effects on coronary blood flow of training consequences which are often opposite in direction is not completely known. Only a very brief account could be considered in this review of such a vast subject, mainly to highlight its perplexing nature and its pertinence to the findings on effects of training.

According to Poiseuille's equation, mean flow could hypothetically be considered to be directly proportional to the perfusion pressure and the 4th power of the radius and inversely proportional to the length of vessel and viscosity of perfusing fluid or blood. At the same level of the latter variables and mean blood flow, values of resistance to the flow may be derived such that they would directly vary with the perfusion pressure required and inversely with the 4th power of the vascular radius. Evidence is available to suggest that coronary blood flow or values of resistance to it is influenced by physical factors related to hydrodynamic variables, properties of the vessel wall and the left ventricle, metabolic factors related to cardiac performance and neurohumoral factors and reflex effects. A small sample of reported reviews (e.g. Gregg and Fisher 1963; Berne and Rubio 1979; Feigl 1983) includes most of these considerations, and a brief description is given below in respect of the coronary circulation with unstenosed vessels.

The physical factors involve extravascular resistance in the myocardium. This is related to ventricular contraction, diastolic length, pressures and ventricular wall tension. The resistance is greater than that related to viscous factors in extramural arteries, such that most of coronary blood flow occurs during ventricular diastole. The vessels are separately subject to elastic wall properties and the myogenic responses to transmural vascular pressure. Though these factors are considered to be relatively small, they have recently been deduced from specially designed studies during changes in perfusion pressure brought about by haemorrhage (Gattullo et al. 1986).

Metabolic factors have been considered to exert a potent vasodilative effect. Mean coronary blood flow and myocardial oxygen consumption are directly proportional to heart rate and ventricular contractile behaviour or pressure generation. Regarding ventricular work, myocardial oxygen consumption is affected to a greater extent by aortic pressure than by cardiac output.

The neural effects have been considered to be relatively smaller than the metabolic ones. These effects involve at least direct sympathetic vasoconstriction and vagal vasodilatation. The reflex effects have been attributed to receptors in the carotid region, coronary vessels and the ventricle.

An intermediary between metabolic and neural effects has been the hydrodynamic factor of perfusion pressure, defined as aortic pressure minus coronary sinus, myocardial tissue or intracavitary pressure. The perfusion pressure would affect flow in the appropriate vascular segments both directly and secondarily through the metabolic effects of afterload and ventricular wall stresses. Blood flow in segments of the coronary circulation within the myocardium is probably marginally greater in the subendocardium than in the subepicardium. In the case of unstenosed coronary vessels under normal experimental conditions, any difference in perfusion related to cardiac haemodynamic performance is not considered great enough to reverse the flow gradient between layers of the myocardium. This issue assumes further relevance in narrowed coronary vessels, as will be mentioned later (Sect. 3.2.2).

Therefore, it is possible to suggest that training-induced decreases in heart rate and increases in left ventricular contractile or pump performance may respectively lead to decreases and increases in coronary blood flow. It could also be argued that through their effect on extravascular coronary resistance and afterload-related metabolic vasodilatation, changes in left ventricular wall thickness, size and intracavitary pressure influence myocardial blood flow. However, of these training-induced changes, the decrease in heart rate has been considered to be the largest single factor influencing coronary blood flow.

Against the above background of interfering concomitant factors, the reports on the effects of training on coronary blood flow will be reviewed in groups, according to the experimental preparations used.

Isolated Perfused Heart. As reviewed in Sect. 3.1.2, it has been possible in isolated perfused heart preparations to control or define variables such as heart rate and aortic and atrial pressures and to exclude reflex effects. Such studies have involved cross-sectional comparsions of hearts between groups of trained and sedentary animals.

In the reports involving male rats (Penpargkul and Scheuer 1970; Scheuer et al. 1974; Bersohn and Scheuer 1977; Giusti et al. 1978; Schaible and Scheuer 1979), coronary perfusate flow was measured by collection, myocardial oxygen consumption was derived by incorporating measured arteriovenous differences and the results were expressed relative to left ventricular dry weight. There was no evidence of ventricular hypertrophy. Hearts from trained and sedentary rats were compared at constant heart rate and mean aortic pressure at set levels of left atrial filling pressure. Trained rats had greater levels of, and increases in, coronary perfusate flow during elevation of atrial pressure. They also had greater coronary arteriovenous oxygen differences and myocardial oxygen consumption, and there were no significant differences in ventricular end-diastolic pressures or volumes. However, the same hearts showed greater ventricular pump and contractile performance at the various atrial pressures. Moreover, these training-related improvements in ventricular performance and coronary perfusate flow were not found following 2 weeks of detraining. Though quantification in such studies is of limited value, it is notable that training-related improvements in myocardial vessels and extracoronary collateral size in male rats were found during detraining to last longer then 2 weeks, as was mentioned in the preceding sections on coronary vascular structure (Leon and Bloor 1968). In these reports, it is difficult unequivocally to rule out the contention that training-related increases in coronary perfusate flow did not correlate with vascular changes, as they could have been related to the concomitant increase in ventricular contractile performance and its secondary metabolic consequences.

Similar preparations were used to examine female rats (Schaible et al. 1981; Schaible and Scheuer 1981). In contrast to male rats, it was reported that training-related improvements in cardiac performance occurred in female hearts only when training-induced hypertrophy was also present, as mentioned in Sect. 3.1.2. Changes in coronary perfusate flow and myocardial oxygen consumption were found to follow the same pattern.

In the reports involving age influence described in Sect. 3.1.2, coronary perfusate flow was also assessed. In the study of Starnes et al. (1983), old trained male rats showed greater peak systolic pressure and cardiac output during high levels of cardiac loading stress than sedentary old rats. Similar trends in average group values were reported for coronary perfusate flow and myocardial oxygen consumption. In the study of Fuller and Nutter (1981), trained young and adult male rats were reported to have greater cardiac output and coronary perfusate flow at high left atrial filling pressures. During recovery from hypoxia, which was examined in the young rats, the left ventricular systolic pressure and coronary perfusate flow in the trained hearts were closer to the previous baseline levels, though group differences relative to sedentary rats did not attain statistical significance (Fuller and Nutter 1981).

In these reports in the isolated perfused heart of the rat, training was associated with a greater coronary perfusate flow and may have been related to greater myocardial contractile performance.

Coronary Perfusion in Isolated Heart. The isolated heart of the rat has also been used during training to assess changes in coronary flow and occurrance of cardiac hypertrophy by utilizing a modified Langendorff retrograde-perfusion apparatus to perfuse the coronary circulation with oxygenated Krebs-Henseleit buffer (Buttrick et al. 1985a, 1985b). Female and male rats were trained by swimming 75 min twice daily 5 days/week, and assessments were made by comparing hearts from matched sedentary groups about six times throughout the training period. Coronary perfusate flow was measured by flow probes and collection at constant pacing heart rate and inflow pressure before and after vasodilataion by anoxic perfusion or addition of adenosine.

In female rats, training resulted in an increase in dry heart weight, which in proportion to initial values was found to begin on the 10th day and progress on the 20th day of training. The only improvement in coronary perfusate flow associated with the training occurred during vasodilatation and was found to occur earlier than the hypertrophy. A similar improvement occurred in trained male rats despite the absence of heart hypertrophy. The findings were considered consistent with those indicating that training-induced changes in coronary vascular structure were independent of heart hypertrophy (Buttrick et al. 1985a). These considerations were supported by findings in a further study using the same preparation in female rat hearts. In comparison with control rats, coronary perfusate flow during vasodilatation was greater in trained rats and smaller in rats with renovascular hypertension. Both interventions resulted in cardiac hypertrophy, and swimming in hypertensive rats improved such coronary flow (Buttrick et al. 1985b).

Whether or not the reported improvements in coronary perfusate flow were influenced by changes in ventricular volume, pressure or stress is not known, since these variables were not assessed.

Anaesthetised Animals. Reports are available which involve training and coronary blood flow in anaesthetised animals. In contrast with isolated heart preparations, changes in haemodynamic variables and their reflex effects occurring concomitantly with training would be expected to influence coronary blood flow.

Findings in the rat have been reported by Wexler and Greenberg (1974), Spear et al. (1978), Yipintsoi et al. (1980) and Koerner and Terjung (1982). In the report of Spear et al. (1978), male rats were trained by running on a treadmill 1 h daily 5 days/week for 12−18 weeks and were compared with matched sedentary groups with or without restriction of food to permit matching of body weight. The trained rats had greater ventricular weights and gastrocnemius cytochrome c concentration. Coronary blood flow was measured using radioactive microspheres before and during hypoxic ventilation with or without increases in aortic pressure brought about by the infusion of methoxamine. Interference by changes in aortic pressure was considered as accounted for by normalising coronary flow to aortic diastolic pressure as the driving pressure during diastole, and the normalised value was labelled as coronary conductance. During hypoxia, trained rats had greater coronary flow and conductance, which were attributed to both training and ventricular hypertrophy. However, average heart rate and cardiac work per minute were also greater in the trained rats, though the diastolic intervals, considered as the "primary period of coronary perfusion", were not significantly different. During hypoxia and raised aortic pressure, trained rats showed better maintenance of coronary flow and greater increases in coronary conductance. However, the trained rats also showed trends towards higher indices of ventricular contractility and cardiac work per minute. These results were considered to indicate training-induced increases in coronary flow and myocardial perfusion in vasodilated coronary circulation, reflecting expected improvements in coronary vascular structure (Spear et al. 1978).

In the report of Yipintsoi et al. (1980), mentioned in Sect. 3.1.2, the effects of hypoxia and volume loading by dextran infusion were examined and myocardial blood flow was measured by the microsphere technique. No significant differences in myocardial blood flow attributable to training were found between trained and sedentary groups of rats. Differences were reported in haemodynamic variables during interventions, such as lower heart rates in trained rats during hypoxia. This report highlighted possible explanations for its findings against the background of reported improvements in coronary structure and flow. The severity of hypoxic intervention and the viscosity of blood compared with that of perfusing fluids in the isolated hearts were considered (Yipintsoi et al. 1980), as alluded to earlier.

In the report of Wexler and Greenberg (1974), young and old male rats were trained by swimming 30 min daily for 2 weeks and compared with sedentary groups. Trained rats had cardiac hypertrophy, and the average heart rate of the group decreased during training. In all the rats, acute myocardial infarction known to heal within days was considered to have been caused by the administration of isoprenaline. Only trained old rats showed benefits in terms of electrocardiographic changes and survival rate, and there were no differences in serum enzyme levels pertaining to the infarction (Wexler and Greenberg 1974). The influence of any concomitant change in haemodynamic variables is not known.

The report of Koerner and Terjung (1982) assessed regional myocardial perfusion using radioactive microspheres. Young male rats were trained by running on a treadmill up to 1 h daily 5 days/week for 12–24 weeks. Ventricular weights did not differ from those in sedentary rats, though cytochrome c levels in the vastus lateralis muscle were greater. Following ligation of the left coronary artery, changes in coronary blood flow were assessed in regions of the left ventricle marked by staining techniques; they included the normally perfused region, the centre of the flow-deprived region and the border region in between. In one series of studies, trained rats were reported to show a trend of greater increases in border region blood flow relative to the normal flow during elevations in aortic diastolic blood pressure over the range of 40–150 mmHg caused by aortic constriction or haemorrhage. These rats, however, maintained lower left ventricular end-diastolic pressures, which could have influenced extravascular coronary resistance. In other series, no differences were found in changes in myocardial blood flow caused by coronary ligation performed during adenosine infusion. The findings were considered to indicate the possibility of a small training-related improvement in myocardial blood flow, which could be obtained in border regions but not in the severely flow-deprived myocardium (Koerner and Terjung 1982). It remains to be determined whether or not such a small improvement was influenced by metabolic and physical factors related to changes concomitant with training in haemodynamic variables or regional myocardial performance.

In the dog, there have been reports which involved assessment of coronary flow during various cardiac stresses (e.g. Laughlin et al. 1978; Bove et al. 1979; Carey et al. 1983). In the report of Laughlin et al. (1978), dogs were trained by running on a treadmill up to 50 min daily 5 times weekly for 10 weeks. In comparison with a sedentary group, the trained dogs had greater heart/body weight ratios and gastrocnemius cytochrome c oxidase activity. The peak reactive hyperaemic flow after 10-s occlusion of the left anterior descending artery was assessed using flowmeters at similar heart rate and arterial blood pressure. This flow was greater in the trained dogs, which also had greater myocardial blood flow. However, it was not established whether this flow improvement was due to larger coronary vasculature or to differences in the operation of vasodilative mechanisms. As assessed with radioactive microspheres, trained dogs had greater myocardial blood flow and a trend towards lower vascular resistance, which was derived by including the mean aortic pressure. During infusion of isoprenaline to increase the heart rate to 200 beats per minute or more, the subendocardial/subepicardial flow ratio was less than unity in the two groups of dogs. Trained dogs, however, maintained a higher ratio, the mechanism of which was not determined (Laughlin et al. 1978).

In the report of Bove et al. (1979), mentioned in Sect. 3.1.2, myocardial blood flow was assessed by the microsphere technique during acute pressure and volume loading. Trained dogs were reported to show significant increases in myocardial blood flow, which did not occur in sedentary dogs. However, the trained dogs had lower baseline myocardial blood flow, and only these dogs showed significant increases in heart rate during pressure loading and in mean aortic pressure during both types of loading.

Carey et al. (1983) assesed myocardial blood flow using radioactive microspheres, during pacing to increase the heart rate to about 200 beats per minute or during adenosine infusion. In contrast to a sedentary group of dogs, another group was trained by running on a treadmill 50 min daily 5 days/week for 8 weeks. The trained dogs showed a decrease in heart rate before and during treadmill exercise tests, and their left ventricular weight was not different from that of the sedentary dogs. In the pacing assessment, no significant differences were found between group data in myocardial blood flow, coronary resistance derived from mean aortic blood pressure alone or in left ventricular oxygen consumption. The average group data for the trained animals showed a tendency towards greater blood flow and oxygen consumption and lower coronary resistance before pacing. The changes with pacing in trained dogs tended towards a greater increase in blood flow or a greater decrease in coronary resistance. Changes in left ventricular dimensions or pressure were not measured, particularly since there were differences between the two groups in cardiac output. In addition, the experiments during adenosine infusion were associated with a greater decrease in average heart rate in the trained dogs, amounting to about 18 beats per minute (Carey et al. 1983).

In another report, pigs were trained by running, mainly on a circular track, up to 1 h daily 5 days/week for about 10 months and were compared with a sedentary group (Sanders et al. 1978). During training, the heart rate decreased before and during exercise, and there was no evidence of cardiac hypertrophy in the trained group. Blood flow in myocardial layers was assessed by the microsphere technique, before and after occlusion of the left circumflex artery and with or without elevation of aortic blood pressure, to examine the effects of raising the perfusion pressure on dilated coronary circulation. No statistically significant differences between the two groups were found in myocardial blood flow or its layer distribution (Sanders et al. 1978). However, trained pigs showed a significant decrease in stroke volume during coronary occlusion and raised aortic pressure; no assessments were made of ventricular pressure, dimensions or border regions.

The review of these reports in anaesthetised animals may be considered to highlight the difficulty of distinguishing changes in myocardial blood flow of regions or layers of the left ventricle attributed to training from those related to changes in the respective perfusion pressures, as well as the interaction of these changes with concomitant differences in ventricular performance, haemodynamic variables and reflex effects.

Coronary Cannulation and Retrograde Flow. There have been reports of studies in anaesthetised dogs which included cannulation of a coronary artery and measurement at its distal segment of blood flow retrogradely arriving from the myocardial vascular bed (Burt and Jackson 1965; Cohen et al. 1978).

In the report of Burt and Jackson (1965), dogs were trained by running on a track up to 90 min daily 5 times per week for 4–6 weeks and were compared with sedentary dogs. The left circumflex artery was ligated and after its distal end had been perfused by blood from the left carotid artery at a pressure of 100 mmHg, retrograde distal flow was measured every other minute. Trained dogs had greater peak retrograde flows, though differences between the two groups did not attain statistical significance. Also, trained dogs maintained a higher flow for a longer period than the sedentary dogs. However, haemodynamic data were not available, and both groups showed electrocardiographic changes indicative of myocardial damage.

In the other report, beagle dogs were trained on a treadmill by alternate days of endurance and sprint running for 10–12 weeks and were compared with a sedentary group (Cohen et al. 1978). Trained dogs showed decreases in heart rate during treadmill exercise tests and higher levels of gastrocnemius cytochrome *c* oxidase activity, but their left ventricular weights were not different from those of sedentary dogs. The left anterior descending artery was perfused by blood from the left carotid artery, myocardial blood flow was measured using radioactive microspheres before and after clamping of the coronary perfusion line, and retrograde coronary flow was collected before and after occlusion of the thoracic aorta. No significant differences were found in myocardial blood flow or its layer distribution, nor in retrograde flow or its conductance in terms of the ratio of flow to mean aortic pressure. However, the cardiac output was significantly greater in the trained group, and it is not possible to rule out an influence of differences in ventricular dimensions or tension or a change in myocardial border regions.

In these reports, retrograde flow was considered to represent an index of collateral blood flow. As will be mentioned later (Sect. 3.2.2), the significance of this flow is debatable, particularly in the case of intact coronary vessels, and it is believed to be subject to the influence of extracoronary vascular resistance, as in the case of myocardial vascular bed.

Coronary Transport in Anaesthetised Animals. There is limited information on the influence of training on transcapillary transport, and two available reports on anaesthetised dogs are considered here (Laughlin and Diana 1975; Laughlin 1985). Training in the two studies included running on a treadmill up to 50 and 75 min daily 5 days/week for 10 and 12–20 weeks respectively. In comparison with sedentary dogs, trained dogs had higher cytochrome *c* oxidase activity, and in one report (Laughlin and Diana 1975), there was evidence of cardiac hypertrophy. In these studies, the left anterior descending coronary artery was perfused through its orifice by blood from the femoral artery, and blood was sampled from the coronary sinus.

In the report of Laughlin and Diana (1975), trained dogs had higher values for coronary blood flow and lower values for resistance. These dogs also showed average trends towards lower aortic pressure. In this study, the fractional extraction of diffusible indicators and the permeability surface area product, which includes fractional extraction and blood flow, did not differ significantly between the two groups of dogs.

In the report of Laughlin (1985), measurements were made before and after the infusion of adenosine, additional prazosin for alpha$_1$ receptor blockade and papaverine into the coronary perfusion line. During constant pressure perfusion, adenosine and prazosin infusion resulted in greater increases in coronary blood flow, as measured by flowmeters in trained dogs, and similar results were obtained in terms of perfusion pressure during constant flow perfusion. There were no significant differences in extraction or permeability surface area product before coronary vasodilatation; after the dilatation and for the same plasma flow, the product was greater in the trained dogs.

These findings were considered to indicate training-related improvements in terms of increases in coronary blood flow and in indices of capillary permeability and area during vasodilatation. The findings also suggested a relationship with other reports on improvements in vascular structure, whether involving available capillary surface area or microvascular pressure (Laughlin 1985). Of pertinence to this review is the possibility raised by such findings of improvements in coronary blood transport to supply the myocardium during conditions of cardiac stress and increases in coronary blood flow, e.g. during exercise.

Conscious Animals. Reports are available of studies on conscious animals, which mainly involve comparisons between trained and sedentary groups (Restorff et al. 1977; Barnard et al. 1980; Breisch et al. 1986). In the report of Restorff et al. (1977), dogs were trained by running on a treadmill in bouts to attain a heart rate of more than 215 beats per minute 80 min daily for 8 weeks. Compared with untrained dogs, no evidence of cardiac hypertrophy was found. During exercise tests, the myocardial blood flow, as measured by radioactive microspheres, increased to a lesser extent in trained than in untrained dogs. This difference was attributed to smaller increases in heart rate and arterial blood pressure in the former group. During exercise, trained dogs had lower mean aortic blood pressure, heart rate and myocardial oxygen consumption, and the ratio of flow in inner to that in outer myocardial layers was always greater than unity; no data were given for the higher levels of exercise. The results indicated changes consistent with those encountered during reductions in myocardial demand for oxygen and in blood flow.

In the report of Barnard et al. (1980), which was mentioned in Sect. 3.1.2, training resulted in left ventricular hypertrophy. Myocardial blood flow was measured using radioactive microspheres. Before and during intermediate levels of exercise, trained dogs had lower left ventricular blood flow, which was associated with a lower heart rate and tension-time index, calculated to represent myocardial demand for oxygen. However, left ventricular diastolic pressure-time index (DPTI), calculated to represent oxygen supply, was greater. These results were considered to indicate a more favourable balance between myocardial demand for, and supply of, oxygen in the trained dogs. During the highest level of exercise, there were no significant differences in left ventricular blood flow or in the calculated indices. However, average group data showed that trained dogs tended to have greater blood flow and tension-time indices, although these did not attain statistical significance. The comparisons respectively involved groups of 14 and 13 untrained and 5 and 4 trained dogs. In the same study, subendocardial resistance during drug-induced vasodilatation was estimated as the ratio of DPTI to subendocardial blood flow; no systematic differences in this resistance between the two groups of dogs were established, despite the occurrence of training-induced hypertrophy. These findings were considered to reflect, before and during exercise, an effect of training-induced reductions in myocardial demand for oxygen, as well as an improvement in terms of subendocardial resistance and oxygen balance to meet training-induced hypertrophy (Barnard et al. 1980). Clearly, these findings raise the possibility, as do other reports reviewed, that the consequences of training-induced effects in terms of reductions in myocardial demand for blood might mask training-induced improvements in coronary blood flow, if any has occurred.

In the report of Breisch et al. (1986), which was mentioned in Sect. 3.2.1, myocardial blood flow in the pig was measured using radioactive microspheres before and during exercise with or without adenosine infusion. Statistically significant differences between trained and untrained groups included greater increases in subepicardial blood flow during exercise in trained pigs, leading to lower ratios of subendocardial to subepicardial flow, particularly during adenosine infusion. These findings were correlated with the structural ones, which indicated reductions in the number of capillaries (Breisch et al. 1986). Other differences in this cross-sectional study included marked training-induced myocardial hypertrophy and lower heart rates during exercise; no assessments were made of changes in cardiac dimensions or pressure.

Longitudinal studies in conscious animals have been reported; one series has involved assessments of phasic coronary blood flow in the dog (Stone 1980b; Liang and Stone 1982, 1983; Gwirtz and Stone 1984; Liang et al. 1984). Training involved running on a treadmill up to 75 min daily, with alternating sprint and endurance periods, 5 days/week for 4–8 weeks or longer, to obtain training-induced reductions in heart rate during exercise testing. Before the occurrence of such bradycardia, dogs were considered partially trained (Stone 1980b). Though studies were carried out before and after training, a sedentary group of dogs were also observed; trained animals had higher skeletal muscle citrate synthase activity, and no changes were reported in variables studied in the sedentary group.

In one study, coronary flow velocity was measured using Doppler ultrasonic probes placed around the left circumflex artery (Stone 1980b). Right atrial pacing was used to increase the heart rate to about 240 beats per minute. By the time of partial training, the dogs had greater increases in coronary blood velocity during pacing, which was maintained to the end of training. No significant changes were reported in left ventricular systolic pressure or its rate of rise. These findings were considered to suggest an early improvement in coronary flow related to an improvement in vascular structure, mainly in the epicardial region.

Peak coronary blood velocity during reactive hyperaemia after 10-s occlusion did not change, at a time when the heart rate and left ventricular pressure were said to have changed very little if at all. Possible changes in wall tension, dimensions or thickness were not ruled out.

During repeated submaximal exercise tests, coronary blood velocity decreased during exercise in the partially trained state, with no changes in heart rate, but eventually showed increases in the trained state towards the baseline untrained values despite decreases in heart rate. Myocardial oxygen consumption was measured in some of these dogs and was found to show trends towards a progressive decrease with training, though statistical significance was not achieved. Similarly, no changes were found in the relationship between myocardial oxygen consumption and heart rate during training, despite expected increases in myocardial contractility and changes in size. Training was found to increase coronary arteriovenous differences, which, with expected changes in ventricular mass (Stone 1980b), raised the possibility of changes related to coronary structure and capillary transport during vasodilatation (Laughlin 1985).

In a subsequent report (Liang and Stone 1982), left circumflex coronary flow was derived using the cross-sectional area of the vessel obtained post mortem, and diastolic coronary resistance was derived from aortic pressure and diastolic coronary blood flow. Studies were performed in the untrained state and following partial training, which was considered to cause an increase in left ventricular end-diastolic volume. Some of the dogs were eventually detrained. During atrial pacing, no changes occurred in aortic diastolic blood pressure, and there were increases in diastolic coronary blood flow and decreases in diastolic coronary resistance. These changes were reversed by detraining and were not found in sedentary dogs. In some dogs, no changes were found in myocardial oxygen consumption or arteriovenous differences. In the same study, beta- or alpha-adrenergic blockade, which resulted in haemodynamic changes, was used to test the proposition that the improvements during pacing were more likely to be associated with coronary structural changes than neural effects.

Similar assessments were made during submaximal exercise testing in another study (Liang and Stone 1983). Improvements were found in diastolic coronary flow and resistance during exercise, as in the pacing study. This improvement during exercise was considered to have been related to coronary vascular structure, as well as training-related reductions in sympathetic vasoconstrictive effects.

A further study (Gwirtz and Stone 1984) was completed in which the effects of pharmacological blocking agents on total coronary blood flow during exercise were examined, following intracoronary administration to avoid large haemodynamic changes. A role was construed for sympathetic vasoconstriction, but not vasodilatation.

In the last study of this series (Liang et al. 1984), myocardial blood flow was measured longitudinally using radioactive microspheres before and during adenosine-induced maximal vasodilatation at a constant heart rate in sedentary, partially trained and trained groups of dogs. Trained dogs had heavier left ventricles, though for groups of five and six dogs the difference was not statistically significant. No statistically significant differences or changes were found in myocardial blood flow in any ventricular layer or in resistance to the flow, which was derived using mean aortic blood pressure. Partially trained dogs were reported to show a tendency towards increased layer blood flow or decreased resistance, and during adenosine infusion, only sedentary dogs showed an average ratio of subendocardial to subepicardial blood flow of less than unity. However, adenosine always caused a decrease in mean aortic blood pressure and presumably in coronary perfusion pressure. Also, maximal vasodilatation was determined by the extent of peak hyperaemic left circumflex artery velocity, which in this series of studies was not changed by training. If the velocity in this artery is felt to represent left ventricular myocardial blood flow (Liang et al. 1984), then it is reasonable to assume that a training-induced change during adenosine infusion is masked by experimental design, as is also supported by the small numbers and the statistically insignificant trends of improvement.

Taken collectively, these five reports raise the possibility that training is associated with an improvement in coronary blood flow, the establishment of which is perhaps made possible by a known baseline pretraining value for the same rather than different animals, and by the use of cardiac stressing in terms of increases in heart rate by pacing or exercise. The reports also highlight the possibility of interference by concomitant haemodynamic changes and reflex effects.

Considering all the reports reviewed in unoccluded coronary vessels, the findings have not shown improvements in coronary flow as consistently as in coronary structure. It is possible to find grounds for the hypothetical proposition that such an outcome was not totally unexpected; factors such as changes in regions or layers of the myocardium and interference by haemodynamic changes are involved.

In respect of coronary structure, stimulation of capillary growth in the myocardium during training could be attributed to the influence of relative hypoxia or metabolites. Such influence is expected to operate in subendocardial layers of the myocardium, at least through vascular compression related to changes in ventricular dimensions. Alternatively, long-term bradycardial pacing has been proposed to result in capillary proliferation (Hudlicka 1982); training-induced bradycardia would lead to capillary growth through longer diastolic periods and distension of the vessels.

The inconsistency in demonstrating increases in blood flow could be due to their magnitude relative to that of the effect of interfering variables and experimental design. The possibility of a small improvement in coronary blood flow has not been unequivocally ruled out. As will be mentioned in the next section, any structure-related improvement in intramyocardial collateral vessels is of a limited nature in terms of functional flow increases, and species differences have been implicated. It is difficult to prove a training-related increase in blood flow which is greater than the effects of changes in cardiac performance, particularly those of the well-established, potent metabolic mechanisms. Some cross-sectional studies of isolated hearts may offer some support in favour of this kind of increase, e.g. by rigid control of variables related to metabolic demands and of the viscosity of the perfusate used. Improvements have also been demonstrated in regions bordering the flow-deprived central region and were small, as expected by virtue of the limited functional capabilities of the collateral circulation. In conscious animals, the reported improvement could be argued to have been small, but unmasked by longitudinal comparisons and the assessment of coronary flow during ventricular diastole.

Findings in a few reports have suggested the possibility of a training-related benefit in terms of increases in transcapillary transport. These issues would assume relevance in the context of effects of training on myocardial ischaemia, as will be considered later in this review.

Narrowed Coronary Vessels

There have been reports that involve training of animals in which the coronary arteries have been narrowed or occluded. Any improvement in myocardial blood flow to flow-deprived regions or layers to be tested would largely depend on the availabilitiy and functional capacity of the appropriate collateral circulation. All the interfering variables encountered in the previous section would complicate the issues of narrowing and collateral circulation, as well as the possible occurrence of variable degrees of myocardial infarction. A very brief outline of these aspects is warranted at this stage.

In hearts with unoccluded coronary vessels, a systematic occurrence of anastomoses between capillaries of two separate major coronary arteries and the ability of collateral vessels to contain a substantial flow have not been established. Following narrowing or slowly progressing occlusion of such arteries, evidence has been reported that collateral vessels may actively grow. In functional terms, the ability of collateral vessels formed in this way to sustain flow rates which significantly compensate for increases in flow expected during augmented performance of the heart has been controversial. Experimental evidence suggests that the ability to increase collateral flow or conductance is limited to about a third of normal values and is further limited by extravascular resistance imposed during increases in heart rate or ventricular pressures. Such limitations are thought particularly likely in at least two myocardial segments. The region bordering the flow-deprived area has been argued to contain a variable number of collateral flow and flow-deprived cells in proportion to the anticipated inadequacy of collaterals. Subendocardial layers of the myocardium are thought more likely to be inadequate in meeting flow increases, and the occurrence of collaterals and their limited flow also vary according to the species of animals. Moreover, collateral flow conductance would be least in infarcted segments of the myocardium (e.g. Schaper et al. 1972; Schaper and Wusten 1979; Flameng et al. 1979; Okun et al. 1979; Bache and Dymek 1981; Newman 1981; Factor et al. 1982; Bache and Schwartz 1983).

From these findings, it could be argued that any training-related increases in coronary blood flow following narrowing or occlusion of the arteries would be small. The improvement in flow in unoccluded coronary beds in such studies was small, and any improvement in the collateral circulation connecting them to flow-deprived beds would occur within the limitations attending it. An improvement in collateral flow would hardly be expected for example in some animal species when two of the three major coronary arteries have been occluded and myocardial infarction and cardiac failure have been induced.

Examples will be considered of reports involving studies of animals or their isolated heart.

Anaesthetised and Conscious Animals. Studies have been reported on the effect of training in animals following narrowing or occlusion of major coronary arteries (Eckstein 1957; Heaton et al. 1978; Neill and Oxendine 1979; Bloor et al. 1984).

In the report of Eckstein (1957) in dogs, several grades of narrowing of the left circumflex artery were imposed. The animals were assigned to a sedentary group or another which was trained by running on a treadmill in four 15- to 20-min sessions daily, 5 days/week for 6–8 weeks. Dogs with gross myocardial infarction were not included, and there were no differences between the two groups in ventricular weight or aortic blood pressure. The aortic blood pressure during anaesthesia was kept at pre-anaesthesia levels, and carotid blood was used to perfuse the distal part of the circumflex artery at the same pressure. Antegrade flow beyond the narrowing was measured by collection and used to assess the narrowing. The retrograde flow from the same artery was collected and its maximal value considered to reflect flow during hypoxia in the collateral circulation. Distal coronary pressure was also measured. In all dogs, greater retrograde flow was found with more severe narrowing, and the flow was higher in the trained dogs. It was also noted that retrograde flow did not develop when narrowing was not severe, though this flow was always present and was greater in the trained dogs (Eckstein 1957). Clearly, the retrograde flow did not represent collateral flow which would exist relative to the resistance in the vascular bed of flow-deprived regions. However, it could be argued that this flow assessment made possible the demonstration of training-related improvements otherwise masked by the restrictions on collateral flow in the intact circulation. The findings of this study were consistent with the minimal extent of collateral flow in the normal heart as reviewed above.

In the report of Heaton et al. (1978) in foxhounds, the left anterior descending artery was occluded by an ameroid constrictor for 3 days and the left circumflex artery narrowed by 60% − 90% of the cross-sectional area. Regional myocardial blood flow was measured using radioactive microspheres during exercise tests, and this assessment was repeated after training by running on a treadmill 1 h daily 5 days/week for 6 weeks. Similar measurements were made in another group of sedentary foxhounds after similar occlusions. Trained dogs developed significantly lower heart rates during exercise, and no scarring was seen in assessed myocardium. In all dogs before training, the normally perfused regions showed increases in myocardial blood flow during exercise in the subendocardial and subepicardial layers. In flow-deprived regions during exercise, flow increases, subendocardial flow and ratios of subendocardial to subepicardial flow were lower, leading to subendocardial underperfusion. These findings were not unexpected, as mentioned above in the review on the behaviour of collateral flow, and there were no differences between the two groups of dogs. In the trained dogs, the only significant change was an improvement in flow during exercise-induced underperfusion in the subendocardial layer of the flow-deprived region, at a time when no such improvement occurred in the sedentary dogs. It was argued that the findings were not causally related to myocardial performance, since improvements were not seen in the normally perfused region, and that the decrease in heart rate or small changes in aortic blood pressure did not correlate with the observed improvement (Heaton et al. 1978). However, in the trained group, trends towards improvement were also apparent in subendocardial layers of normally perfused regions but did not attain statistical significance, and the average heart rates during exercise were 184 and 163 beats per minute before and after training respectively. Moreover, changes in flow assessed by the microsphere technique include those which occur in both normal and collateral beds. This issue was examined in another study, in which coronary blood flow in the myocardium and retrograde flow were assessed in the same animals, as will be mentioned below.

In the report of Neill and Oxendine (1979), dogs with occlusion of the left circumflex coronary artery, which was caused over 2–3 weeks by ameroid constrictors, were assigned either to sedentary groups or to groups trained by running on a treadmill up to 30 min daily 5 days/week for 5 or 8 weeks. Only dogs trained for 8 weeks had a decrease in heart rate during submaximal exercise, though these dogs did not show evidence of cardiac hypertrophy and their left ventricular volume and ejection fraction, as determined by left ventriculography in the conscious state, were not different from those of sedentary dogs. Myocardial scarring was excluded from analysis. Studies were performed during atrial pacing-induced tachycardia of up to 250 beats per minute in the 5-week-trained, and up to 200 beats per minute in the 8-week-trained dogs. In the former group, pacing led to a decrease in myocardial blood flow in the flow-deprived region relative to that of the normally perfused one. In both groups, the reduction during pacing in flow ratios of subendocardial to subepicardial layers was greater in the flow-deprived than in normally perfused regions. In contrast, there were no differences between trained and sedentary groups. Left coronary angiography with dogs under anaesthesia was considered to be probably too insensitive to detect changes in coronary vasculature. Retrograde flow was greater in trained than in sedentary dogs and tended to be further increased by the longer training time of 8 weeks (Neill and Oxendine 1979). This study again demonstrates the occurrence of underperfusion during cardiac stressing. Also, the findings were consistent with the contention that any change in collateral flow is subject to limitations attributable to resistance to blood flow in the myocardium and to the tendency in the dog for collateral vessels to develop mainly in the subepicardial layers of the myocardium, though their full development is thought to require longer than the period of this study (e.g. Schaper et al. 1972).

A report is available (Bloor et al. 1984) of a study in pigs in which collaterals are thought to develop in the subendocardium, mainly in the papillary muscles and the interventricular septum (Schaper et al. 1972), and to be sparser than in the dog (Bloor et al. 1984). The left circumflex artery in pigs was narrowed to reduce reactive hyperaemia to 15% of its prestenotic value (Bloor et al. 1984). Four groups were studied, which comprised sedentary sham-operated, trained intact, sedentary and trained coronary narrowing groups. Training involved running on a treadmill at 70%–100% of the maximal heart rate up to 45 min daily 5 days/week for 5 months. This was associated with decreases in heart rate during exercise tests. Regional myocardial blood flow was measured using radioactive microspheres, at similar mean aortic blood pressure, before and after release of occlusions in each of the three coronary arteries. Collateral flow was greater in pigs with narrowing and even more so in the trained animals with narrowing, particularly in the regions bordering the centre of the flow-deprived region. However, the collateral flow was always less than normal coronary flow. Anatomical studies showed complete occlusion of the circumflex artery and the occurrence of myocardial infarction, though the area of infarct laterally was less in the trained pigs. Of interest was the finding that scar tissue contained collateral flow (Bloor et al. 1984). It is possible to construe that a benefit in terms of tissue salvage could have occurred, though it remains to be established whether the difference in collateral flow occurs independently of the scar size and resistance to blood flow, which in addition could have involved differences in ventricular dimensions. In the absence of such knowledge or of changes in flow in myocardial layers, acceptance of an improvement in collateral flow, believed to be meagre in the pig, would assume relevance.

Isolated Heart. Even meticulous attempts in the isolated heart to test training-induced effects on the collateral circulation have yielded opposite conclusions (Scheel et al. 1981; Schaper 1982). In the first report (Scheel et al. 1981), four groups of beagles were studied, comprising a group of sedentary animals, a group trained by running on a treadmill 45 min daily 5 days/week for 6 weeks, a sedentary group undergoing ameroid constrictor occlusion of the left circumflex coronary artery and, finally, a group with occlusion which subsequently underwent training for 8 weeks. Trained beagles were not considered to have developed cardiac hypertrophy, but their heart rate was lower following training. In an isolated beating heart preparation, the three major coronary arteries were cannulated and perfused with blood at a preset constant pressure. Coronary blood flow was measured using flow probes, and experiments were completed during adenosine-induced vasodilatation at constant perfusion pressure and with the ventricles vented to atmospheric pressure to minimise changes in afterload. The resistance to coronary flow in each of the three arteries was derived by relating perfusion pressures to late diastolic coronary flows. No statistically significant differences between trained and sedentary groups were found, though the trained dogs tended to have lower coronary resistance. The collateral resistance between various combinations of coronary beds was assessed by retrograde flow measurement with a prefusion pressure of 100 mmHg. Trained dogs with occlusion had lower resistances between the left anterior descending, septal or right coronary arteries and the occluded left circumflex artery. These findings indicated a training-related improvement in collateral circulation, which was considered to include epicardial and intramyocardial collateral vessels by virtue of septal artery circulation (Scheel et al. 1981). These findings are consistent with the hypothetical contentions outlined in this review that changes in retrograde flow could have been unmasked be removing the limiting effect of coronary bed resistance, as expected in the intact heart.

The finding of a wider collateral circulation extending deeper than the subepicardial layer of the beagles contrasts with that found in other dogs, as has been previously observed (Schaper et al. 1972). It should be noted, however, that the report of Scheel et al. (1981) did not rule out the probability that the greater coronary resistance to antegrade flow found in sedentary dogs with only one occluded vessel was attributable to scarring.

In the second report (Schaper 1982) on trained dogs, two coronary arteries, the left circumflex and right coronary, were occluded for 2.5 weeks by ameroid constrictors, and postoperatively a group of trained dogs was compared with a sedentary group. Training involved running on a treadmill 1 h daily 5 days/week for 12 weeks. Trained dogs developed slower heart rates during exercise, and their cardiac weight was slightly greater than in the sedentary group. The hearts were excised, connected to a Langendorff apparatus and perfused with blood from support dogs. Regional myocardial blood flow was measured by radioactive microspheres, and experiments were performed during adenosine-induced vasodilatation at various levels of perfusing pressure. No significant improvements were shown in the relation of perfusion pressure to layer flow with or without consideration of the distal coronary pressure. Also, even in such a rigidly controlled preparation, it was notable that the increase in collateral flow during vasodilatation was substantially less than the increase in normal coronary blood flow (Schaper 1982).

Regarding this condition of underperfusion with occlusion of two of the three major coronary arteries, it should be pointed out that questions arise concerning the reliance on one remaining vessel and the possibility of variable degrees of scarring. The latter could be argued to have occurred during training exercises leading to underperfusion and to have masked any training-related increases in conductance, which are expected to be small. As Schaper (1982) points out, any exercise-induced underperfusion and ischaemia which could stimulate growth of collaterals would be limited to the subendocardium in dogs, which are known to have the capability of developing collateral vessels in the subepicardium.

These findings in animals with narrowed coronary arteries were not entirely unexpected, as pointed out earlier in this section. In a simple approach to the problem, any training-related increase in collateral flow, which would be in series with normal flow and subject to resistive components in both vascular segments, should be small. An improvement would hardly be expected if the source of flow to the collateral vessels were drastically curtailed because of extravascular resistance. Also, in layers of the myocardium, because of a species-related lack of collaterals, a predominant improvement is unlikely. Improvements demonstrated in terms of retrograde flow are perhaps related to minimising effects of masking variables. If is occurs at all in the intact heart, a small change would be difficult to demonstrate. Benefits of limiting tissue death could then suggest different mechanisms, one of which, the possibility of improved capillary transport, has already been raised earlier in this review.

3.3 Summary

The review of reports in experimental preparations has indicated that exercise training results in changes in cardiac performance and coronary circulation. In general, these changes are related to the intensity and duration of training, sex, age and animal species. In the intact animal, demonstration of single training-related changes is influenced by concomitant factors.

There is evidence that training results in a slowing of heart rate, ventricular hypertrophy and possibly a small improvement in cardiac pump performance and myocardial inotropism. Changes in ventricular diastolic volume are variable and, as is the case with cardiac performance or coronary blood flow, are known to be sensitive to changes in haemodynamic variables which include the heart rate. Increases in cardiac output are reported in conditions of cardiac stressing, including exercise.

Training probably results in improvements in coronary vascular structure, particularly increases in the number of myocardial vessels. Despite variability in histological techniques, the improvements are demonstrated with remarkable consistency.

The functional significance of structural improvements is less certain; for example, discernible increases in blood flow in the myocardium are not consistently found. However, the possibility has not yet been unequivocally ruled out that during training flow increases occur in magnitudes relatively smaller than and thus overshadowed by metabolic, physical and neural consequences of concomitant haemodynamic and cardiac changes.

When coronary vessels are occluded and collateral vessels develop subject to a species-determined influence, the function of the latter depends on certain limitations. Assuming an in-series connection to coronary vessels, collateral flow is known to be limited and is vulnerable to extracoronary compressions, as well as to coronary vascular resistance in general. Differences between species such as the sparse development and distribution of collateral vessels have been postulated. These vessels develop in the subepicardium in the dog and to a variable extent in further layers of the myocardium in the beagle, pig and man. It is not surprising that any training-related improvements in collateral circulation, which may be demonstrated within rigid experimental conditions, should have a minimal impact in the intact heart or animal.

Some findings suggest that functional improvement could still occur, e.g. in terms of improved capillary transport during increased performance of the heart and a related increase in coronary blood flow.

It would be difficult meaningfully to quantify a cardiac effect directly in relation to training in general. Cardiac training effects are apparently influenced by the programmes used, as well as by a variable, though definite, interplay between these effects. It is reasonable to assume that intrinsic cardiac effects are small, in that they could be masked by concomitant effects of training, but become more apparent under conditions of cardiac stressing and increased performance.

4 Import of Experimental Evidence in General

As reviewed in preceding sections, reported findings in experimental animal preparations have included evidence for the occurrence of cardiac training effects, as well as factors which could influence and even mask the manifestation of these effects, particularly in intact and conscious animals. A substantial proportion of these findings could be considered in connection with the issue of training in man, in whom, additionally, the influence of other factors is expected. Such factors include at least characteristics of study populations and genetic effects.

This section briefly highlights the importance for man of experimentally demonstrated evidence on cardiac training effects. Clearly, most of the techniques which reliably control the influence of interfering variables in experimental animal preparations cannot be used in man. However, the demonstration of similar cardiac training effects in man would at least raise the probability of fundamentally common backgrounds. The evidence thus obtained will be relevant to Sects. 5 and 6 of this review, which involve recent techniques, namely exercise tests, developed to assess the effect of training on cardiorespiratory fitness and ischaemic heart disease in man.

4.1 Cardiac Performance

Trials in man have demonstrated general and cardiac performance effects of training which were similar to those found in experimental animals; these effects have been described in detail (e.g. Astrand and Rodahl 1977; Clausen 1977; Schaible and Scheuer 1985; Cox et al. 1986) and will be briefly outlined in this section.

A decrease in heart rate has been a consistent effect of training, particularly during submaximal exercise. Other training-related changes demonstrated have included an increase in left ventricular diastolic wall thickness and dimensions, which should be considered against the background of a decrease in heart rate and prolongation of ventricular filling period. Increases in blood volume have also been reported.

In respect of cardiac performance, training has been reported to result in increases in ventricular stroke volume at rest and during exercise and an increase in cardiac output during maximal exercise. An increase in oxygen consumption during maximal exercise has been shown to result from training, and was related to increases in cardiac output and skeletal muscle adaptations.

It has not been unequivocally demonstrated whether changes occur in the inotropic state and the Starling mechanism of the ventricle. This demonstration would at least make it possible to attribute the increases in ventricular stroke volume and cardiac output during exercise to intrinsic cardiac mechanisms. Assessment of such ventricular performance during exercise in healthy subjects, is difficult. The use of recent methods such as echocardiography or radionuclide techniques is subject to certain limitations (e.g. Gibson 1984; Wackers 1984), which assume relevance in detecting small changes. Studies using these techniques have yielded inconsistent results (e.g. Schaible and Scheuer 1985). Technical difficulties include achievement of adequate records, assumptions related to geometrical dimensions or background stability and the fact that these techniques assess changes in dimensions during ventricular ejection. Moreover, in intact subjects any assessment of changes in intrinsic ventricular performance during exercise would be limited by the interference of many variables, which include heart rate, ventricular loading and reflex mechanisms (e.g. Mary 1986).

It could be concluded in general that results of training in man have been shown to be essentially similar to those in animals, as reviewed in preceding sections. In respect of mechanisms in man, it is not unreasonable to assume that the training-related increases in ventricular output have included small increases in pump performance and contractility.

The decrease in heart rate during submaximal exercise and the increase in oxygen consumption during maximal exercise have been used as indices of the effectiveness of training programmes. The two variables are dependent on the intensity and the duration of training, as is the case in reports in animals. As briefly outlined in this section, the effects in man have been demonstrated after training which was sufficient to influence the two indices. It is as yet unknown whether training programmes which are not sufficient to change these indices would consistently result in effects which could be attributed to the heart. This issue will assume relevance in the context of cardiorespiratory fitness, to be reviewed in Sect. 5.

4.2 Coronary Heart Disease

In subjects without coronary artery disease, there has not been an adequate number of studies to allow unequivocal conclusions on effects of training on the coronary circulation. In a cross-sectional study, a lower coronary sinus flow was found in trained volunteers, which was associated with lower heart rate levels and tension-time index (e.g. Stone 1980a; Schaible and Scheuer 1985).

In the pathological condition of coronary heart disease, some effects of training, which include decreases in heart rate during exercise, are qualitatively similar to those in healthy subjects. In coronary heart disease, however, maximal oxygen consumption may not be attained during exercise and left ventricular dimensions are likely to increase to a greater extent than in the normal case (e.g. Clausen 1976). As outlined in Sect. 4.1, in the case of healthy subjects, it is difficult to attribute changes in myocardial performance to training. Further interfering factors include the extent of myocardial damage related to coronary artery disease, its marked variability between patients and the difficulty of imposing adequate cardiac stress. All these complications could be implicated in the inconsistency of reported changes in myocardial performance (e.g. Sim and Neill 1974; Kennedy et al. 1976; Letac et al. 1977; Nolewajka et al. 1979; Froelicher et al. 1980; Hagberg et al. 1983; Ehsani et al. 1986).

In respect of the effects of training on the coronary circulation in patients with coronary heart disease, certain issues need to be considered. As was reviewed in Sect. 3.2.2, it would be difficult to distinguish a direct effect of training from those which concomitantly influence coronary blood flow. This is apparent even in studies in experimental animal preparations, where more accurate techniques of assessement can be used and less variability is encountered than in patients with coronary artery disease. The disease occurs in patients under examination with a variable extent of severity, progression or myocardial damage and is clearly different pathologically from experimental narrowing or occlusion. The latter issues make it necessary to interpret the findings in the context of considerations such as the adequacy of the examined sample to represent the disease in general, the influence of training on the pathology of the disease and the appropriate duration required to manifest such an influence. Examples of some reports will be outlined to highlight such considerations.

Coronary blood flow has been examined by methods which may be considered to assess flow directly, though they cannot measure regional perfusion in myocardial layers. The methods have included clearance techniques during atrial pacing to increase the heart rate (Sim and Neill 1974), thermodilution (Ferguson et al. 1978), thallium-201 scintigraphy (e.g. Scheuer 1982) and radioactive macroaggregated albumin (Nolewajka et al. 1979) during exercise.

In the report of Sim and Neill (1974), no changes were found during training, and the technique was considered mainly to reflect flow in normally perfused myocardium. In the eight patients examined, coronary artery disease varied markedly in severity, and coronary flow either increased or decreased (Sim and Neill 1974). Similar variability in coronary artery disease and coronary sinus flow or myocardial perfusion during exercise have been reported (Ferguson et al. 1978; Scheuer 1982; Lynch and Crawford 1983). In the report of Ferguson et al. (1978), coronary sinus flow was mainly influenced by concomitant haemodynamic changes. In the report of Nolewajka et al. (1979), coronary artery disease progressed in patients following myocardial infarction whether trained or not. Assessment of myocardial capillary flow, which was not considered accurate in reflecting regional variations, did not suggest an improvement. These reports are consistent with the contention, based on findings in experimental animal preparations, that it would be difficult to demonstrate small improvements in coronary flow in the face of wide variability in the severity of disease, technical limitations and interference by haemodynamic changes.

In respect of coronary artery disease, examples will be considered of reports which raise the possibility of training effects on the pathological occlusive process. Repeated coronary angiography has been used to assess changes in vascular dimensions or in the development of collateral vessels. The findings, which comprise inconsistent changes, have previously been reviewed (e.g. Hellerstein 1969; Froelicher et al. 1980; Scheuer 1982; Wyatt 1982). However, it is not certain whether or not collateral vessels could be measured using angiography, and inconsistency is known to occur in measurements of coronary narrowing. Furthermore, the relationship between the degree of narrowing and coronary blood flow is not linear (e.g. Linden and Mary 1982). It has been pointed out that such techniques are of insufficient accuracy in quantitative assessments, and the time required for the occurrence of structural changes could be too long for the period of study, particularly against the background of interpatient variation in disease severity or rate of progression (e.g. Ferguson et al. 1974; Nolewajka et al. 1979; Scheuer 1982).

Regarding the pathology of coronary artery disease, reports are available which raise the possibility of training benefits. Monkeys were trained before and during 2 years' consumption of atherogenic diets to raise the level of serum cholesterol and were compared with sedentary groups (Kramsch et al. 1981). Trained animals had cardiac hypertrophy and developed slower heart rates at rest and during exercise. They were reported to have smaller lesion size and less collagen accumulation, as assessed by angiography and postmortem histological examinations. These animals also had greater high-density lipoprpotein cholesterol and lower triglyceride or low-density lipoprotein levels. Whether or not such induced lesions are the same as in coronary artery disease in man, there have been reports which suggest that training leads to similar changes in plasma catecholamine and serum lipid levels, as well as to increases in fibrinolytic activity (e.g. Froelicher et al. 1980; Rigotti et al. 1983; Ehsani et al. 1984; Rauramaa et al. 1984).

The above review on the coronary circulation bears a resemblance to experimental coronary artery narrowing or occlusion. As reviewed in Sect. 3.2.2, it has been difficult to exclude interference by metabolic, physical and neural consequences of training-related haemodynamic effects. In respect of training benefits, there remains the possibility of changes in transcapillary transport, coronary artery disease or associated pathological processes. These issues will assume relevance in Sect. 6, which considers potential benefits of training on myocardial ischaemia related to coronary artery disease.

5 Assessment of Fitness Effects of Training

Evidence was presented in preceding sections to suggest that training leads to improvement in cardiac performance, particularly during stresses of exercise. Many indices of physical fitness, obtained by non-invasive means, have been associated with changes in cardiac performance. In this section, well-known indices will be briefly reviewed and emphasis placed on recently developed methods of assessment.

5.1 Definitions

Physical fitness in general includes important attributes of the body which involve issues such as skills, body weight, cardiorespiratory function, muscle strength and endurance. Training-related improvement in cardiorespiratory fitness or oxygen transport and utilisation has been associated with changes in cardiac frequency and pumping performance. In the absence of limitations related to ventilation and diffusion, oxygen consumption during exercise would depend on the magnitude of changes in cardiac output and the adequacy of perfusion of organs involved in the exercise (e.g. Holmgren 1967; Sinning 1975; Astrand 1976; Clausen 1977; Bassey and Fentem 1981).

In the assessment of cardiorespiratory fitness, it is pertinent to this review to highlight certain considerations. Within the context of such fitness, an interplay between cardiac and other effects of training is expected. Specifically, changes in cardiorespiratory fitness during exercise would include both cardiac and peripheral adaptations (e.g. Astrand 1976; Clausen 1977), which possibly lead to increases in oxygen extraction (Astrand and Rodahl 1977). Furthermore, the evidence reviewed in preceding sections has indicated that training leads to small changes in cardiac performance which are influenced by the intensity, frequency and duration of the training programme. Therefore, the design of tests for changes in cardiorespiratory fitness would require adequate sensitivity to detect small changes which could reasonably be attributed to cardiac adaptations.

5.2 Methods of Assessment

Cardiorespiratory fitness has primarily been assessed during exercise stressing, and formal exercise tests which could be administered non-invasively have been developed.

Most of the types or designs of exercise tests used have previously been reviewed in detail (e.g. Wydenham 1967; Astrand 1976; Shephard 1978; Bassey and Fentem 1981). In general, indices of cardiorespiratory fitness have involved maximal exercise testing and, less commonly, assessments during submaximal exercise.

5.3 Maximal Oxygen Consumption

Maximal oxygen consumption (VO_2max) has been extensively used as an index of cardiorespiratory fitness. Details of its use and accuracy are available in several adequate reviews (e.g. Holmgren 1967; Pollock 1973; Astrand 1976; Clausen 1977; Scheuer and Tipton 1977; Shephard 1978; Bassey and Fentem 1981), and some aspects will be outlined below.

Increases in VO_2max could be shown to occur during training and have been attributed at least in part to increases in cardiac output during maximal exercise. However, VO_2max is known to be influenced by several factors other than cardiorespiratory fitness.

Measurement of oxygen consumption requires apparatus to allow assessments during all steps of work increments in exercise testing, and portable methods have been available. The criteria of maximal exercise used to obtain VO_2max are not always attainable with some individuals or during exercise tests which involve small muscular mass. Maintenance of a plateau of oxygen consumption during maximal exercise requires exhaustive effort which could not be achieved in some sedentary subjects, patients or children. Blood lactate level, respiratory exchange ratio and maximal heart rate have been used in addition. Clearly, premature stopping of exercise tests would yield values of oxygen consumption which are less than VO_2max. Though such values assume relevance in respect of exercise tolerance, as will be mentioned in Sect. 6, they are not thought solely to represent cardiorespiratory fitness.

The value of VO_2max in healthy subjects has been found to depend on conditions which include age, sex and genetic constitution. Progressive decreases occur after the age of 25 years, and higher values have been observed in males than in females. During training, increases in VO$_2$max appeared to be determined by initial levels of fitness, as well as by the intensity, duration and frequency of exercise training. Greater increases have been shown to occur in individuals with low pretraining VO$_2$max than in those with higher levels. In gerneral, systematic increases in VO$_2$max during training have been reported to occur with exercise intensity at greater than 50% – 60% of VO$_2$max, with an exercise duration of about 20 min daily and a frequency of three times per week. However, the effects of such training criteria have varied between individuals and were subject to interactions between the criteria themselves and interferences by initial levels of VO$_2$max, age and muscles involved in the exercise.

The reported magnitude of training-related increases in VO$_2$max in sedentary subjects has varied in relation to the factors mentioned above. Mean increases in groups of subjects have been of the order of 15% – 20%. Such modest increases are not inconsistent with the evidence reviewed in preceding sections that training-related improvements in cardiocirculatory performance are small.

The findings on VO$_2$max could now be considered in relation to its accuracy as an index of cardiorespiratory fitness. Large variability is expected between individuals, and VO$_2$max in one exercise test does not reliably predict the 'level' of cardiorespiratory fitness. In a "homogeneous, clinically healthy and relatively well-conditioned" population, the 95% confidence limits amounted to about 40% (Astrand 1976). As would be expected therefore, VO$_2$max would represent an index of change in, rather than level of, fitness; in this context, reproducibility in the measurement of this index would assume relevance. Differences between repeatedly measured VO$_2$max in the same subjects have been reported; 95% of individual differences, i.e. 95% tolerance limits, were estimated to be about 3.6% – 14% of mean VO$_2$max (Newell 1982). Since VO$_2$max varies directly with the intensity of training, such reproducibility estimates would clearly place limitations on the sensitivity of this index to detect small improvement in cardiorespiratory fitness in the individual. This limitation would add to the other interfering factors pointed out earlier in this section.

It could be concluded from this brief review that VO$_2$max is an useful indicator of changes in cardiorespiratory fitness during strenuous and exhaustive exercise. Its use is limited by the ability to attain maximal levels of exercise, which could depend on factors other than cardiorespiratory fitness. Furthermore, in the individual its use is limited by the wide tolerance limits of measurement relative to expected changes. However, VO$_2$max has been useful in groups of healthy populations subjected to training of sufficient severity.

5.4 Other Indices

To avoid the difficulties encountered in its measurement, various methods have been used indirectly to estimate VO_2max, particularly with submaximal exercise.

Commonly used methods have included linear extrapolation to an assumed age-related maximal heart rate of the relationship between measured heart rate and oxygen consumption throughout various numbers of steps of submaximal exercise. However, such methods suffer from the drawback that they assume levels of maximal heart rate known to vary widely, as well as the existence of a linear relationship between heart rate and oxygen consumption at high levels of exercise, which is not an uniform finding (e.g. Astrand 1976). The variability of such estimates has been documented in a review of international experience (Shephard 1978).

Other indices have been assessed, especially in comparison with VO_2max. Examples of these include the work rate at a set heart rate, heart rate or blood lactate concentration at a set rate of oxygen consumption or work rate, ratio of maximal work rate to heart rate, changes in respiratory exchange ratio and oxygen consumption at a respiratory exchange ratio of unity and combinations of various criteria of exercise performance and anthropometric data (e.g. Roskamm 1967; Wydenham 1967; Astrand 1976; Shephard 1978; Bassey and Fentem 1981; Mortimer and Reed 1982; Weller et al. 1985). These indices were in general reported to correlate in trend with VO_2max and to differ from its individual values. In the context of training-related changes in cardiorespiratory fitness, the accuracy of these indices has not been fully quantified, though potentially they could be useful in patients who are unable to attain maximal levels of exercise.

5.5 Submaximal Heart Rate-Oxygen Consumption Relationship

The use of submaximal exercise to derive indices of cardiorespiratory fitness would obviate the problems of measurement during maximal exercise. As was reviewed in preceding sections, training has been shown consistently to result in decreases in heart rate and increases in ventricular stroke volume during submaximal exercise. The reported changes in cardiac output have not been consistent. During such exercise, oxygen consumption would represent the product of cardiac output, or heart rate times stroke volume, and arteriovenous oxygen difference.

The relationship during submaximal exercise between work rate or oxygen consumption and heart rate or cardiac output could be computed to fit linear regression lines (e.g. Holmgren 1967; Astrand 1976; Bassey and Fentem 1981). Measurement of heart rate, work rate and oxygen consumption may be obtained by non-invasive means.

5.5.1 Assessments in General

As reviewed in preceding sections, training which leads to improvement in cardiac performance also results in decreases in heart rate at submaximal levels of exercise work rate and oxygen consumption. Expressed in terms of computed linear regression lines of heart rate on oxygen consumption, training would be expected to shift the regression line to the right, such that a higher oxygen consumption would be attained at the same heart rate. Examples of such findings will be cited below because of their relevance to the heart rate-oxygen consumption index to be mentioned in following sections.

Reports have been available on the effect of bed rest and training. For instance, in a longitudinal study of young subjects, exercise testing was performed on a treadmill and oxygen consumption measured at submaximal loads at 40%, 60% and 80% of VO_2max (Saltin et al. 1968). Relative to initial control values, mean VO_2max decreased during bed rest and increased during training. No significant change in oxygen consumption was reported at these loads. Mean heart rates at rest and at the given levels of oxygen consumption increased after bed rest and decreased after training (Saltin et al. 1968).

In other studies, middle aged men were examined before and after bed rest (Convertino et al. 1982; Hung et al. 1983). Oxygen consumption was measured at rest and during the last 30 s of four 3-min stages of upright exercise on a bicycle ergometer; the fourth stage represented maximal work. Bed rest was associated with increases in mean heart rate at all levels of oxygen consumption during exercise (Hung et al. 1983). This finding could be construed to represent a shift to the left in the relationship during exercise in mean group values of heart rate on oxygen consumption. In the same subjects, VO_2max decreased and maximal heart rate increased, and it was reported that during rest and all stages of work rate mean oxygen consumption decreased and heart rate increased at rest and all stages of work rate (Convertino et al. 1982). It was notable in this study following bed rest, that increases in mean oxygen consumption at higher work rates were less than those at low work rates, thus suggesting changes in the slope as well as shifts in the level of mean oxygen consumption-work rate relationship.

The computed linear regression relationship of heart rate to oxygen consumption, as measured in the steady state during levels of submaximal exercise, has been used to assess cardiorespiratory fitness. For instance, the slope of the relationship has been directly calculated (Spiro et al. 1974), and a value of heart rate interpolated at a preset level of oxygen consumption was used to reflect the level of the relationship (e.g. Bassey and Fentem 1981). As pointed out above, however, in an adequate analysis of the relationship of heart rate to oxygen consumption, interpolation of single values would be undermined by changes in both the level and the slope of this relation (Bassey and Fentem 1981).

The design of the exercise test is expected to influence the relationship of heart rate to oxygen consumption and work rate. For instance, the reports of Convertino et al. (1982) and Hung et al. (1983) involved comparisons between supine and upright bicycle exercise tests. During the former test, bed rest resulted in increases in mean heart rate only at high levels of oxygen consumption or work rate. In general, bed rest-related decreases in oxygen consumption and increases in heart rate were greater during upright than during supine exercise. Relative to mean values of oxygen consumption, mean heart rates were greater during upright than supine exercise (Convertino et al. 1982; Hung et al. 1983). Similarly, in trained healthy subjects, VO_2max achieved during supine bicycle exercise is lower than during upright bicycle exercise (e.g. Astrand 1976). These findings entail the possibility that haemodynamic and reflex adaptations related to posture could influence the nature of, and changes in, the relationship between heart rate and oxygen consumption during submaximal increments in work rate.

Other findings suggest that heart rates at given submaximal oxygen consumption levels would be greater during bicycle than during treadmill exercise (e.g. Hermansen et al. 1970), and small differences may occur in these values in individuals when obtained during the steady state of discontinuous and continuous increments in work rate (e.g. Bassey and Fentem 1981).

Other examples involve findings of the effects of training the arms or legs, as assessed by exercising either set of limbs (e.g. Clausen 1977). During exercise tests, measurements were made at two work rates after $4-7$ min of exercise. Briefly, leg training resulted in decreases in heart rate relative to oxygen consumption values obtained during leg and arm exercise tests. The resting values were also included to yield three points of relationship of heart rate to oxygen consumption. The decrease in heart rate during untrained muscle exercise tests followed the change which occurred during rest without changing the slope of the relationship. During trained muscle exercise tests, there was an additional reduction in the slope, such that increases in heart rate during high submaximal oxygen consumption were lower than those at lower oxygen consumption (Clausen 1977). It is expected that exercising muscle mass will affect the relation of heart rate to oxygen consumption (e.g. Lewis et al. 1983). Analysis of covariance showed that the slope of the linear computed relationship was steeper with arm than with leg muscle exercise. Maximal heart rate and oxygen consumption were greater during leg exercise, leading to greater heart rates at 50% or more of exercise-related VO_2max during leg than during arm exercise. It is notable that in the same report no differences were found in the computed relationship of cardiac output to oxygen consumption.

It is noteworthy that the findings reviewed primarily apply to healthy subjects, since pathological conditions or drugs could have their own effects on the relationship of heart rate to oxygen consumption and work rate. For instance, healthy volunteers were made to breathe two levels of carbon monox-

ide and examined during three stages of bicycle exercise, including maximal
levels (Ekblom and Huot 1972). Progressive increases in mean values of heart
rate at similar oxygen consumption and progressive decreases in mean
VO_2max were reported. In another report, the effect of reductions in red cell
mass has been examined longitudinally in healthy volunteers (Woodson et al.
1978). Changes in the computed linear relationship of heart rate to oxygen
consumption (HR/VO_2) during supine exercise were assessed by analysis of
covariance. The intervention was shown to result in a significant shift to the
left in the elevation of HR/VO_2. A similar shift was found in the relation of
cardiac output to VO_2, and there was a decrease in mean VO_2max during
treadmill exercise (Woodson et al. 1978).

Acute administration of the beta-adrenergic blocking agent propranolol
has been found, during bicycle exercise, to decrease the heart rate at given submaximal levels of oxygen consumption, as expressed in proportion to
VO_2max attained during treadmill exercise, to reduce cardiac output and the
slope of heart rate relative to oxygen consumption. Opposite changes in levels
or slopes were obtained with atropine (e.g. Astrand 1976). Similar findings
were reported in respect of another beta-adrenergic blocking agent, atenolol
(Hespel et al. 1986). During submaximal upright exercise, oxygen consumption was continuously calculated, and heart rates obtained at levels of oxygen
consumption in increments of 10% of the measured maximal value were used
for comparison with placebo. The resting and submaximal heart rates were
lower with atenolol, and the decrease was greater at high levels of percentage
oxygen consumption, as assessed by three-way analysis of variance.

As in the case of VO_2max, therefore, it is possible from this review to find
evidence in healthy subjects to suggest that changes in heart rate relative to
oxygen consumption and work rate during submaximal exercise might be
related to changes in cardiorespiratory fitness. This relationship could be influenced by variables which possibly include blood volumes, posture, the
mass of muscle involved in training and exercise tests.

5.5.2 Index of Cardiorespiratory Fitness

The relation of submaximal HR to VO_2 has been used to develop an index
of changes in cardiorespiratory fitness in individual subjects (Newell 1982).
Aspects utilised include suitable training programmes and a detailed analysis
of HR/VO_2 throughout submaximal levels of exercise testing.

Training was administered using the Royal Canadian Air Force programme
for men (5BX) and women (XBX). These programmes are defined according
to age in terms of duration, frequency and rate of progression and require
$11-12$ min of exercise daily without the need for specialised equipment.
These features and the nature of training, which mainly involve callisthenics
and a stationary run, were considered to impose only minimal demands on

normal daily activity and thus to be suited for systematically testing changes in fitness. The programme 5BX has been found in a group of soldiers to result in a mean increase in VO_2max (Malhotra et al. 1973).

The exercise test involved cycling in the upright position at submaximal work rates which were presented in pairs in a discontinuous but increasing series. Measurements for analysis were made in the steady state when oxygen consumption and heart rate varied by less than 5% and 2% respectively for at least 1 min. In each exercise test, regression analysis of heart rate on oxygen consumption (HR/VO_2) was performed to yield a linear regression line. Changes in the slope and elevation of these lines in the same individuals were examined for statistical significance using analysis of covariance. In tests repeated on consecutive days, no significant changes were reported in the slope, and those regarding elevation were not consistent (Kappagoda et al. 1979). Seven 'sedentary'subjects were tested during training and after attaining the target levels of the training programme. Training consistently led to a significant shift to the right in HR/VO_2 elevation and reversal after detraining. The only significant change in the slope was a decrease in one subject at the end of training. In contrast, these consistent changes were not obtained when tests were repeated without training after a period equivalent to the length of the training programme. It was notable that the shift in elevation began even before target levels of training were attained. The elevation of HR/VO_2 was considered a valid index for detecting sequential changes in cardiorespiratory fitness in individual subjects (Kappagoda et al. 1979).

This index was used in subsequent studies in healthy volunteers to assess the effect on cardiorespiratory fitness of changes in the components or frequency of the 5BX/XBX programmes (Mary et al. 1986). Omission of press-ups for 8 weeks of daily 5BX training did not significantly affect HR/VO_2. Volunteers who improved in fitness following 5BX/XBX training were serially examined after 4–8 weeks of training twice weekly, daily training, training on alternate days and, finally, daily training. The sequence was reversed for alternate subjects (Fig. 1). These trials showed that improved cardiorespiratory fitness in some subjects could be maintained by training on alternate days or twice weekly (Fig. 2). Any loss of improvement occurred within 4 weeks of twice-weekly training or within 8 weeks of training on alternate days.

The effect of the 5BX/XBX programmes was assessed using the same index of cardiorespiratory fitness in patients recovering from replacement of a single heart valve (Newell et al. 1980). In control patients followed up for 24 weeks without training, variable changes in the elevation of the HR/VO_2 relationship were found. In particular, patients with rheumatic heart disease failed to show shifts to the right in this elevation of HR/VO_2 relationship. In contrast, during training all patients with and without rheumatic heart disease showed a consistent shift to the right in HR/VO_2 elevation. In the same study (Newell et al. 1980) and during subsequent observations of serial test-

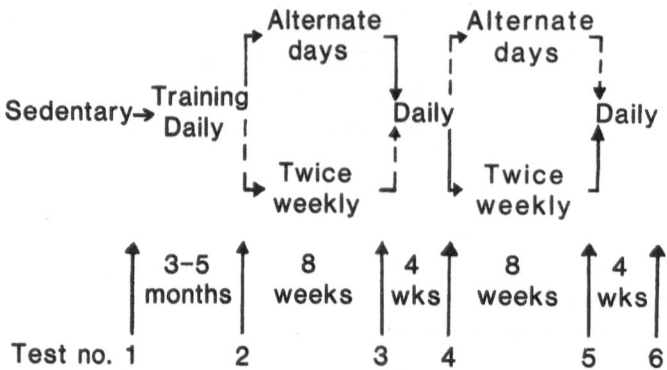

Fig. 1. Trial design of the effects of frequency of training on cardiorespiratory fitness. The *top* part represents the two modes (*continuous* and *interrupted arrows*) of changes in training in alternate subjects. The periods of training and exercise tests undergone by each subject are indicated in the *bottom* part of the diagram

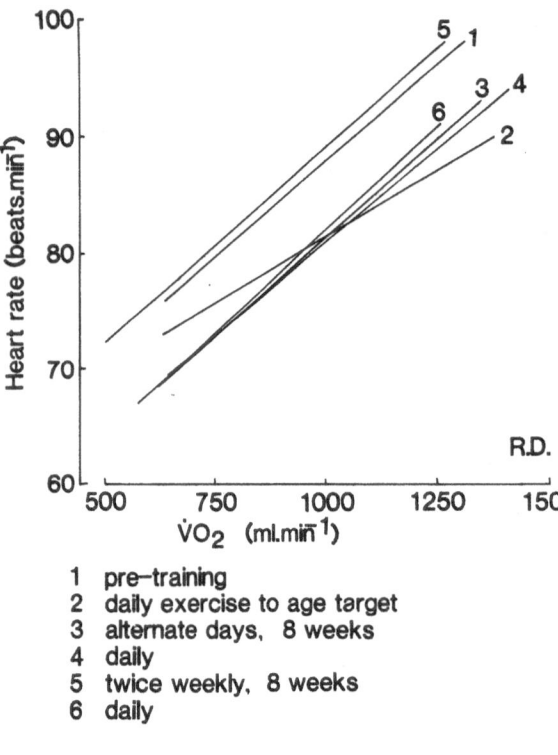

1 pre-training
2 daily exercise to age target
3 alternate days, 8 weeks
4 daily
5 twice weekly, 8 weeks
6 daily

Fig. 2. Example of results obtained in one subject throughout the period of trial. The six *continuous lines* represent the linear relationship between heart rate and oxygen consumption (VO_2) during six exercise tests, their sequence is indicated by the *numbers* and explained at the *bottom* of the figure (see also Fig. 1, *continuous arrows*). A significant shift in the relationship to the right occurred following training daily (*2*) and on alternate days (*3*). The reverse occurred following twice weekly training (*5*)

ings in such patients (Winter, Mary, Ionescu and Linden; unpublished observations), a shift to the left in HR/VO_2 elevation was found in association with development of subsequently diagnosed subacute bacterial endocarditis and heart failure.

In another trial, the effect of indoramin, an alpha$_1$-adrenoceptor antagonist, on the HR/VO_2 relationship was examined in patients with untreated mild hypertension (Bishop et al. 1986 c). In patients in whom there were reductions in arterial blood pressure during exercise and possibly reductions in cardiac afterload, a shift to the right occurred in the elevation of HR/VO_2 relationship. Indoramin also resulted in small decreases in heart rate at submaximal work rates.

These findings demonstrate the potential of the submaximal HR/VO_2 relationship as an index of cardiorespiratory fitness in individual subjects. The mechanisms involved have not been examined; given the findings mentioned in this review, it is not possible hypothetically to rule out an improvement in cardiac pumping performance in relation to oxygen consumption. Further studies are required to quantify the extent of this improvement.

6 Assessment of Training Effects on Ischaemic Heart Disease

In patients with coronary heart disease, the effects of training on myocardial ischaemia have been assessed by methods which mainly involved exercise testing. During such assessments, non-invasive methods have been considered to offer the particular advantage of serial examinations in the same individuals. However, the methods used to detect myocardial ischaemia or its change in relation to training have varied in their accuracy. This section briefly considers the accuracy of commonly used methods of non-invasively assessing ischaemia in relation to training and places emphasis on two recently developed indices of myocardial ischaemia.

6.1 Definitions

It is pertinent to this review to present a summary of commonly accepted definitions of myocardial ischaemia, particularly in relation to exercise in man. The precise mechanisms involved in the process of ischaemia are not completely known, as has previously been reviewed (e.g. Maseri 1975; Hearse 1979; Hoffman 1981).

In coronary heart disease, a simplified and pragmatic definition of myocardial ischaemia provoked by exercise has evolved, involving the failure of coronary blood supply to meet increased myocardial demand for oxygen. This issue has previously been reviewed by Linden and Mary (1982). Though other possible variables including local metabolites or substrates other than oxygen could follow, ischaemia would basically reflect reductions in myocardial oxygen tension, to which subendocardial layers consistently appear most vulnerable. According to these definitions, factors related to myocardial oxygen consumption and supply would be involved in the precipitation of myocardial ischaemia during exercise. Myocardial oxygen consumption per minute, or the product of myocardial blood flow and arteriovenous oxygen difference, is known to be mainly influenced by the heart rate and ventricular contractile behaviour or pressure generation. At constant arterial oxygen content, the supply of blood and oxygen to the myocardium could be reduced by luminal narrowing or spasm of the coronary vessels. In addition, such reductions are expected during short diastolic periods and increases in myocardial wall tension and ventricular pressure, particularly in subendocardial layers of the myocardium (see also Sect. 3.2.2).

6.2 Methods of Assessment

In general, myocardial ischaemia has been assessed by the use of various indicators. During cardiac stressing and increases in heart rate by exercise or atrial pacing, the indices have included occurrence of anginal pain, area of arterial diastolic pressure, electrocardiographic waves, myocardial lactate production and function and segmental perfusion of the left ventricle (e.g. Hellerstein et al. 1965; Detry and Bruce 1971; Redwood et al. 1972; Sim and Neill 1974; Clausen 1976; Barnard et al. 1977; Wallace et al. 1978; Ferguson et al. 1978; Lee et al. 1979; Nolewajka et al. 1979; Froelicher et al. 1980; Raffo et al. 1980; Ehsani et al. 1981; Elamin et al. 1983; Rigotti et al. 1983; Lynch and Crawford 1983).

The considerations regarding myocardial ischaemia mentioned in Sect. 6.1 have basically been involved in the majority of reported indices. For instance, indices of coronary blood supply in terms of myocardial perfusion, arterial diastolic pressure-time area, function or segmental wall movement have been related to indices of myocardial oxygen consumption in terms of heart rate, systolic blood pressure level with or without its duration and ejection time. Also, the indices of myocardial oxygen consumption have been related to indices of myocardial ischaemia in terms of anginal pain, lactate production and electrocardiographic signs.

The relationship between the indices of myocardial ischaemia used and the reviewed effects of training assumes particular relevance. For instance, assessments of transmural coronary blood flow or myocardial perfusion would differ from those of myocardial ischaemia, which is believed to occur predominantly in subendocardial layers. Also, as was mentioned in preceding sections, there were indications that training could involve factors related to myocardial ischaemia other than coronary blood flow or myocardial oxygen consumption, e.g. transcapillary transport. Clearly, the use of indices in the non-invasive assessment of myocardial ischaemia in the same individuals is fundamentally determined by their accuracy in detecting ischaemia or its small changes.

6.3 Exercise Testing

Extensive experience in exercise testing has provided the basis for non-invasive assessments of the effect of training on ischaemic heart disease. Such assessments have in the main involved two interrelated aspects: exercise tolerance, in terms of incremental work rate before limitations are imposed by myocardial ischaemia, and severity of ischaemia, in terms of levels of myocardial oxygen consumption attained during myocardial ischaemia (e.g. Hellerstein et al. 1965; Detry and Bruce 1971; Redwood et al. 1972; Clausen 1976; Froelicher et al. 1980; Raffo et al. 1980; Ehsani et al. 1981; Elamin et al. 1983; Rigotti et al. 1983). This section will consider these aspects in relation to reports involving non-invasive assessments of the effect of training on myocardial ischaemia.

6.3.1 Exercise Tolerance

The fact that training in patients with coronary heart disease makes possible longer duration and higher levels of exercise before the occurrence of anginal pain or myocardial ischaemia has been consistently reported (e.g. Redwood et al. 1972; Sim and Neill 1974; Adams et al. 1974; Clausen 1976; Greenberg et al. 1979; Raffo et al. 1980; Rigotti et al. 1983). Such an improvement has been considered consistent with training-related reductions during submaximal exercise in indices of myocardial oxygen consumption. The latter have included the heart rate, product of heart rate and systolic or mean arterial blood pressure and the product of heart rate, systolic blood pressure and systolic ejection time.

In these reports, it was possible to measure indices of myocardial oxygen consumption objectively. However, their change was related to the occurrence of anginal pain or indices of ischaemia. Clearly, such changes do not objectively confirm or rule out changes in the severity of myocardial ischaemia. As has been pointed out by Froelicher (1973) and reviewed in preceding sections, training-related changes in ventricular contractile performance or dimensions could influence myocardial oxygen consumption and coronary blood supply. The objectivity and accuracy of assessment of angina or ischaemia will be considered in the following section.

6.3.2 ST-Segment Depression

In patients with coronary heart disease, exercise has been shown to result in various electrocardiographic changes. A considerable body of evidence has been available that ST-segment depression could be used in formal exercise testing as an index of the occurrence and severity of myocardial ischaemia (e.g. Linden and Mary 1982; Froelicher 1983).

The mechanisms responsible for the displacement of the ST segment in man have not been unequivocally established. However, evidence has been reported indicating that the magnitude of ST-segment depression is related to that of restriction in the increase of subendocardial coronary blood flow which occurs in relation to increased heart rate (e.g. Linden and Mary 1982; Mirvis and Gordey 1983; Lee et al. 1986; Mirvis et al. 1986). In general, a correlation has been found between the restrictions in myocardial blood flow, intramyocardial oxygen or carbon dioxide tension and ST-segment displacement or decline in regional myocardial contractile performance.

As would be expected according to the evidence reviewed so far, increases in indices of myocardial oxygen consumption during exercise in such patients would lead to increases in ST-segment depression (e.g. Detry and Bruce 1971; Raffo et al. 1980; Linden and Mary 1982). Experimental evidence has also been consistent with these findings. During occlusion of one major coronary vessel and increases in heart rate, the magnitude of ST-segment depression was shown to be directly related to the severity of myocardial ischaemia (e.g. Linden and Mary 1982; Mirvis et al. 1986). These relationships have been used to assess the effect of training on myocardial ischaemia during formal exercise testing in patients with coronary heart disease.

ST-Segment and Heart Rate

The changes in indices of myocardial oxygen consumption attained at the onset of angina or in ST-segment depression during exercise testing have been used as indicators of changes in myocardial ischaemia during training in patients with coronary heart disease; these studies have been reviewed elsewhere (e.g. Adams et al. 1974; Clausen 1976; Greenberg et al. 1979; Rigotti et al. 1983). In such studies, myocardial oxygen consumption was assessed using heart rate with or without arterial blood pressure or systolic ejection time. As indicated in this review, increases in these variables would imply increases in the supply of oxygen to the myocardium and therefore either reductions in the severity of myocardial ischaemia or reductions in other variables related to myocardial oxygen consumption, such as left ventricular contractile performance or dimensions. These studies have not consistently shown that training results in increases in the indices of myocardial oxygen consumption used.

It is possible to suggest that such inconsistency could be related to variability in the methods used to select reference levels of severity of myocardial ischaemia at which the haemodynamic variables were assessed. For instance, the use of anginal pain as an index of myocardial ischaemia is limited by its subjective nature and vulnerability to the influence of factors other than ischaemia and by changes reported during training in the relationship of pain to ST-segment depression (e.g. Raffo et al. 1980; Linden and Mary 1982). In respect of the use of ST-segment depression, its magnitude could increase during training in association with increases in work rates and heart rate, as mentioned earlier in this section and reported by Detry and Bruce (1971). Similarly, training-related decreases in heart rate at submaximal work rate would lead during training to the development of ST-segment depression at higher submaximal work rates, as was shown by Nolewajka et al. (1979). During such assessments, decreases or increases respectively in submaximal or 'maximal' ST-segment depression could involve improvements related to exercise tolerance, as was mentioned in Sect. 6.3.1, as well as those in ischaemia.

More recently, Raffo et al. (1980) used a modified exercise test to assess the effect of moderate training on myocardial ischaemia. The test was developed to relate levels of heart rate and systemic blood pressure to an objectively determined level of myocardial ischaemia, as assessed by the occurrence during steady-state exercise of ST-segment depression of 0.1 mV. The test comprised two parts of exercise on a bicycle ergometer; during the first part, the work rate was increased every 3 min until the occurrence of ST-segment depression of at least 0.1 mV in electrocardiographic recordings of a bipolar lead CM_s. During the second part, smaller step increments in work rate were used to increase the heart rate in smaller steps than in the first part and to permit measurements in the steady state in terms of heart rate and ST-segment depression of 0.1 mV. The latter heart rate (HR) was labelled HR/ST threshold (Fig. 3) and used as an objective index of changes in the severity of myocardial ischaemia.

In reproducibility studies, the 95% tolerance limits of single differences of HR/ST threshold from the mean amounted to 2.5 beats per minute (Raffo et

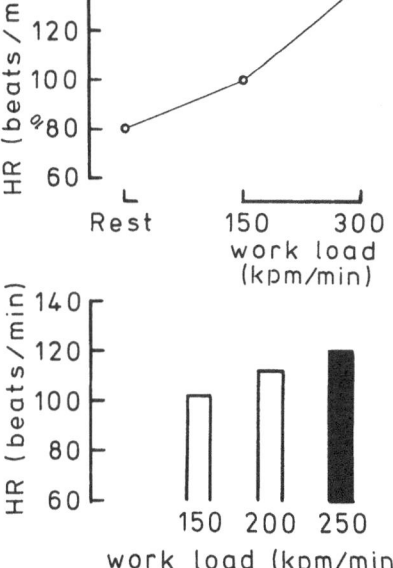

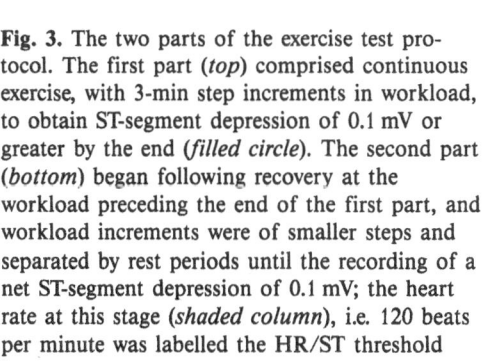

Fig. 3. The two parts of the exercise test protocol. The first part (*top*) comprised continuous exercise, with 3-min step increments in workload, to obtain ST-segment depression of 0.1 mV or greater by the end (*filled circle*). The second part (*bottom*) began following recovery at the workload preceding the end of the first part, and workload increments were of smaller steps and separated by rest periods until the recording of a net ST-segment depression of 0.1 mV; the heart rate at this stage (*shaded column*), i.e. 120 beats per minute was labelled the HR/ST threshold

al. 1980). In a separate study, an increase in HR/ST threshold was shown to occur in patients who continued to have ST-segment depression following aortocoronary bypass (Luksic et al. 1981).

The effect of a moderate training programme (5BX/XBX) on HR/ST threshold was examined in patients with stable angina pectoris due to coronary heart disease who had ST-segment depression during exercise (Raffo et al. 1980). The patients were randomised into two groups, a training and a control group. As was mentioned in Sect. 5.5.2, this training programme has been shown to improve cardiorespiratory fitness in healthy volunteers and in patients following heart valve replacement. In contrast to the control group, training was shown to result in reductions in the heart rate during submaximal work loads (Raffo et al. 1980). In preliminary studies in such patients, the relationship of heart rate to work load was compared with that of heart rate to oxygen consumption (Winter, Mary and Linden; unpublished observations). Following training, significant decreases in the elevation of the heart rate/work load relationship, as assessed by analysis of covariance, were always associated with significant decreases in the elevation of the heart rate/oxygen consumption relationship. However, in individual patients, decreases in the elevation of the heart rate/work load relationship which did not reach statistical significance could be associated with a significant reduction in the elevation of the heart rate/oxygen consumption relationship. As mentioned in Sect. 5.5.2, these data objectively indicated training-related improvements in cardiorespiratory fitness in patients with coronary heart disease (Raffo et al. 1980).

Patients in the training group had an increase in HR/ST threshold, and those in the control group had a decrease in this threshold. Changes in systolic blood pressure were not significant, leading to similar results in terms of heart rate and systolic blood pressure product/ST threshold (Raffo et al. 1980). These results were considered to indicate reductions in the severity of myocardial ischaemia following training, possibly through the ability to attain greater levels of myocardial oxygen consumption at similar severity of ischaemia, as objectively assessed by the same level of ST-segment depression in the same patients. The heart rate alone provides a simple and accurate measurement during such non-invasive exercise tests and, as outlined earlier in this review, has a direct relationship with increases in myocardial oxygen consumption during exercise. Over a period of time, the heart rate would provide an integral of developed ventricular wall tension and pressure generation. Furthermore, training-related changes in heart rate are more significant than in blood pressure, and in general, experimental evidence has suggested that reductions in cardiac dimensions or contractile performance are not as likely or as consistent as the decrease in heart rate. These considerations were supported by the small variability in the measurement of the HR/ST threshold and its diametrically opposite changes respectively in randomised groups of patients with and without moderate training (Raffo et al. 1980).

During a follow-up study, which partly involved patients from the trial of Raffo et al. (1980), the same methods were used to examine the effect of long-term maintenance training of up to 4.5 years (Winter et al. 1984). Similar improvement, indicating a reduction in the severity of myocardial ischaemia, was shown to be sustained or increased in patients with coronary heart disease whether or not they were receiving beta-blocking agents. This study confirmed the results obtained by Raffo et al. (1980), in a further population of patients, of changes in myocardial ischaemia. In one patient who stopped training, there was a decrease in the HR/ST threshold which occurred in parallel with the absence of a significant change in the heart rate/work load relationship. Also, in two patients who maintained improvement in cardiospiratory fitness, there was a decrease in the HR/ST threshold, indicating progression of myocardial ischaemia.

A similar approach to that of Raffo et al. (1980) has been reported, though a different training programme and exercise test were used (e.g. Ehsani et al. 1981). In this study, patients with a history of myocardial infarction or with hyperlipoproteinaemia and who did not have angina pectoris were examined during intensive training and compared with a control group. The heart rate and systolic blood pressure were measured at the time during exercise when ST-segment depression of 0.1 mV first appeared in three consecutive electrocardiographic cycles. In each patient, training resulted in increases in maximal oxygen consumption and heart rate at the target ST-segment depression. An increase in systolic blood pressure occurred less consistently than did an increase in the heart rate, and in the whole group the increases in heart rate, systolic blood pressure and their product at the target ST-segment depression were statistically significant. In the same patients during three levels of submaximal exercise, ST-segment depression was progressively less during increases in double product. Also, there was an increase in echocardiographically measured end-diastolic left ventricular dimensions. Subsequently, using the same training programme Ehsani et al. (1986) examined patients with coronary heart disease who, during exercise, had no change in ejection fraction or developed segmental left ventricular wall motion abnormalities. Left ventricular function and volumes were assessed during rest and supine exercise using electrocardiographically gated cardiac pool imaging. Briefly, training in one group of patients resulted, during supine exercise, in smaller ST-segment depression and greater left ventricular ejection fraction and end-diastolic volume at equivalent double product. Ehsani et al. (1981, 1986) considered these findings to indicate a training-related decrease in the severity of myocardial ischaemia, for reasons similar to those mentioned above in the report of Raffo et al. (1980).

The findings reviewed suggest that exercise testing trials could be designed to deduce training-related reductions in the severity of myocardial ischaemia. However, it would be difficult to quantify the extent of such reductions. The findings also showed differences between patients in indices of myocardial oxygen consumption at similar levels of ST-segment depression. This issue will be considered in the next section, which involves the use of a quantitative index of myocardial ischaemia.

Maximal ST-Segment/Heart Rate Slope

Whilst exercise electrocardiography tests have been used over the past few decades in the assessment of myocardial ischaemia in patients with coronary heart disease, they have also been shown to suffer from drawbacks and limitations to their accuracy. This subject has been reviewed in detail elsewhere (e.g. Linden and Mary 1982), and only a brief account could be given in this review.

A substantial proportion of exercise tests have been designed to rely, either solely or amongst other variables, on ST-segment depression during or after exercise. Qualitatively, a preset level of depression has been used to detect occurrence of myocardial ischaemia, and quantitatively, the extent of ST-segment depression has been used to detect its severity. The reliability of such tests has mainly been determined by trials in hospital patients with angina who have undergone the exercise test and coronary angiography. In this population, it is widely reported that such exercise tests may fail to detect the presence or absence of myocardial ischaemia and its severity. Possible explanations that have been suggested include variability between patients, e.g. in the ability to exercise, ventricular wall abnormalities, size and pressure. As was reviewed in earlier sections, these variables could influence myocardial blood flow. In respect of the ability to exercise, the measurement of ST-segment depression is objective, but its occurrence at the end of exercise is subjectively determined and often depends on factors other than myocardial ischaemia. Indeed, the exercise tests have continually been subjected to modifications, which have included the type of exercise, electrocardiographic leads and the setting of derived indices.

Recently, an exercise test was developed to avoid the variability of obtaining indices at subjectively determined events in time, such as the end of exercise, and to include factors which enhance the accuracy of the test by relating indices of myocardial oxygen consumption to ST-segment depression throughout the exercise test (e.g. Linden and Mary 1982). Briefly, the precision of measuring the heart rate in the non-invasive setting and its relationship to myocardial oxygen consumption during exercise, as mentioned in earlier sections of this review, were considered, and ST-segment depression was measured with the aid of a magnifying glass. There were indications in groups of patients that the level of heart rate considered in combination with the extent of ST-segment depression was related to the severity of myocardial ischaemia. In particular, experience from longitudinal studies (Raffo et al. 1980; Luksic

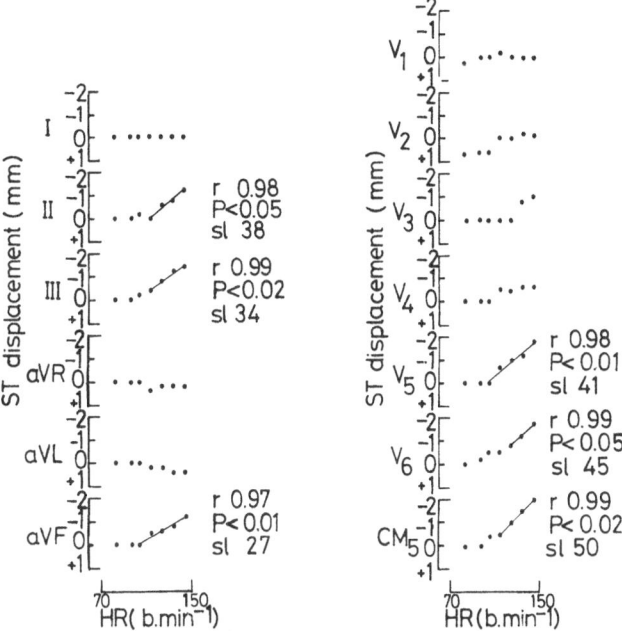

Fig. 4. The results of the maximal ST/HR slope test for a patient with angina and narrowing of two major coronary arteries. Each *dot* represents the average of measurements in at least ten consecutive cardiac cycles recorded during the steady state of each step of exercise on the 13 electrocardiographic leads. The test was tailored to the patient by a preliminary exercise to effect step increases in heart rate of about 10 beats per minute. Regression analysis in each lead was performed (Linden and Mary 1982) to obtain linear lines (*continuous lines*) with the steepest slopes (*sl*) of the relationship of ST-segment depression on heart rate (*HR*). The maximal ST/HR slope was the greatest of all the slopes obtained in all leads, i.e. $50 \, mm \cdot beats^{-1} \cdot min \cdot 10^{-3}$ in lead CM_5

et al. 1981; Ehsani et al. 1981), as reviewed in the preceding section, indicated that the slope of heart rate or double product relative to ST-segment depression could be related to the severity of myocardial ischaemia, since changes were observed in the thresholds at a time when the baseline values were similar.

Following preliminary trials in learning populations, the relationship between heart rate and ST-segment depression throughout exercise was found to be linear, as derived from a specially designed exercise test which is tailored for individual patients. The maximal slope derived from standard 12 electrocardiographic leads and a bipolar lead CM_s (maximal ST/HR slope) was used as an index of myocardial ischaemia (Fig. 4). In reproducibility studies, the 95% tolerance limits of individual differences between repeated measurement of the maximal ST/HR slope amounted to 1.9%.

The evidence for the accuracy of the slope in detecting the presence of myocardial ischaemia and quantifying its severity in terms of coronary

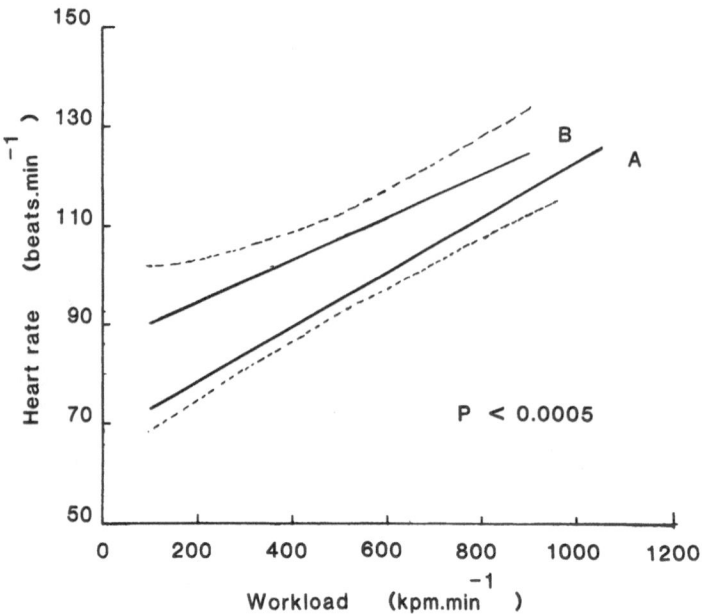

Fig. 5. Pooled data for heart rate and workload during exercise tests for eight patients, before (*B*) and after (*A*) training. The *continuous* and *interrupted lines* represent the computed regression relationship between the two sets of data and their 95% confidence limits respectively. There is a statistically significant shift to the right in the elevation of the relationship with training

angiographic changes in patients with coronary heart disease has been reviewed in detail elsewhere (Linden and Mary 1982; Bishop et al. 1987). Briefly, using 95% confidence limits of the binomial distribution, the slope detected myocardial ischaemia attributable to coronary artery disease in 96% of selected patients with angina. The disease was assessed by independent analysis of coronary arteriograms. Only visual assessment and team judgement were used to avoid error of measurement, and severe narrowing of approximately greater than 75% stenosis of proximal parts of major coronary vessels was used to ensure certainty of ischaemic heart disease in the presence of symptoms. The maximal ST/HR slope was not significantly affected by beta-blocker therapy and was found reliable in studies before and after coronary angioplasty or aortocoronary bypass to quantify severity of ischaemia in terms of number of narrowed coronary vessels.

In patients with angina, the accuracy of the maximal ST/HR slope may be limited by the presence of left ventricular enlargement, impaired function or wall scarring, and in the presence of conduction defects such as right bundle branch block and accelerated conduction. In general, however, the slope was found more accurate in detecting myocardial ischaemia and its severity than other tests in use (Linden and Mary 1982; Bishop et al. 1987). It should be

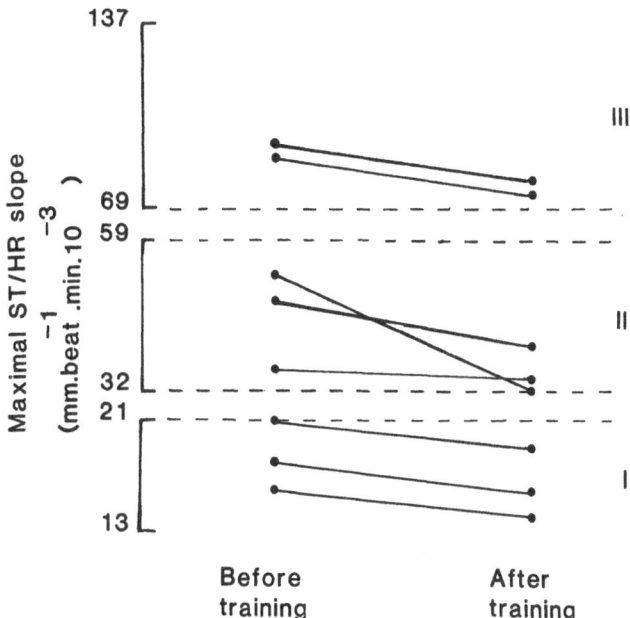

Fig. 6. The maximal ST/HR slope before and after training for the eight patients described in Fig. 5. The limits of ranges of the slope in patients with angina according to the number of narrowed major coronary arteries (*I, II* and *III*) (Linden and Mary 1982). With training, a reduction occurred in the maximal ST/HR slope; each reduction was within the limits of severity of coronary artery disease

pointed out that there have been reports which could not confirm the accuracy of the slope. Reports have been available which discuss possible explanations for differences in findings (Kligfield et al. 1986; Bishop et al. 1987) and present mounting evidence of the superior accuracy of ST/HR slopes as indices of myocardial ischaemia (Okin et al. 1986; Finkelhor et al. 1986; Kligfield et al. 1985, 1986; Bishop et al. 1987). The ST/HR slope has also been used in patients with angina to examine the effects of various vascular dilating agents on the severity of myocardial ischaemia, which are separate from those related to changes in exercise tolerance as defined in Sect. 6.3.1 (Bishop et al. 1986a, 1986b; Berkenboom et al. 1986).

The maximal ST/HR slope was used to examine the effect of the training programme 5BX/XBX, mentioned earlier in this review, on myocardial ischaemia in patients with coronary heart disease (Elamin et al. 1983). In some patients examined, the slope accurately detected the severity of myocardial ischaemia in terms of number of narrowed major coronary arteries. Following training, the patients developed significant reductions in heart rates at submaximal exercise workloads (Fig. 5); with each patient, there was a reduction in the maximal ST/HR slope (Fig. 6). The findings were considered to indicate a reduction in the severity of myocardial ischaemia. Changes in heart rate or blood pressure have not been found to affect the maximal ST/HR slope, including those induced by pharmacological beta-blockade (e.g. Linden and Mary 1982; Okin et al. 1985). Despite these reductions in the severity of myocardial ischaemia, the patients retained values of the slope indicating the same number of narrowed coronary vessels. However, the training programme was of moderate intensity, and it remains to be demonstrated whether a more intensive programme would induce further changes.

The findings reviewed in this section have shown that exercise tests carefully designed to provide sensitive indices of myocardial ischaemia and serial assessments in individual patients indicate the possibility that training results in reductions in the severity of myocardial ischaemia.

The mechanisms of this improvement remain unknown. However, the evidence given earlier in this review could be used hypothetically to propose possible improvements involving the supply of blood and oxygen to subendocardial layers of the myocardium. With the expected limitations of collateral vessels and increased myocardial dimensions, any increase in blood flow related to coronary structural changes or neural effects would be small and inconsistent. However, evidence was presented to suggest improvement in transcapillary transport and in coronary arteriovenous oxygen difference. The latter, rather than the uncertainty of collateral circulation, would assume relevance in attributing the reduction in severity of myocardial ischaemia to regional improvement in oxygen availability. It is remarkable that indices of myocardial ischaemia, in concert with the reviewed evidence, could only suggest small improvements, which were perhaps limited to regions of the myocardium.

References

Abbott BC, Mommaerts WFHM (1959) A study of inotropic mechanisms in the papillary muscle preparation. J Gen Physiol 42:533–551

Adams WC, McHenry MM, Bernauer EM (1974) Long-term physiologic adaptations to exercise with special reference to performance and cardiorespiratory function in health and disease. Am J Cardiol 33:765–775

Allen DG (1983) The use of isolated cardiac muscle preparation. In: Techniques in the life sciences, vol. P3/1. Elsevier Scientific Publishers Ireland Ltd. Linden RJ (ed). Cardiovasc Physiol P310:1–21

Amsterdam EA, Choquet Y, Segel L, Arbogast R, Rendig S, Zelis R, Mason DT (1973) Effects of exercise conditioning on the rat heart: physical, metabolic and functional correlates. Circulation 48 [Suppl IV]:137

Anversa P, Beghi C, Levicky V, McDonald SL, Kikkawa Y (1982) Morphometry of right ventricular hypertrophy induced by strenuous exercise in rat. Am J Physiol 243:H856–H861

Anversa P, Levicky V, Beghi C, McDonald SL, Kikkawa Y (1983) Morphometry of exercise-induced right ventricular hypertrophy in the rat. Circ Res 52:57–64

Astrand P-O (1976) Quantification of exercise capability and evaluation of physical capacity in man. Prog Cardiovasc Dis 19:51–67

Astrand P-O, Rodahl K (1977) Physical training. In: Astrand P-O, Rodahl K (eds) Textbook of work physiology. McGraw-Hill, New York, pp 389–445

Bache RJ, Dymek DJ (1981) Local and regional regulation of coronary vascular tone. Prog Cardiovasc Dis 24:191–212

Bache RJ, Schwartz JS (1983) Myocardial blood flow during exercise after gradual coronary occlusion in the dog. Am J Physiol 245:H131–H138

Baker MA, Horvath SM (1964) Influence of water temperature on oxygen uptake by swimming rats. J Appl Physiol 19:1215–1218

Baldwin KM, Cooke DA, Cheadle WG (1977) Time course adaptations in cardiac and skeletal muscle to different running programs. J Appl Physiol 42:267–272

Barnard RJ, MacAlpin R, Kattus AA, Buckberg GD (1977) Effect of training on myocardial oxygen supply/demand balance. Circulation 56:289–291

Barnard RJ, Duncan HW, Baldwin KM, Grimditch G, Buckberg GD (1980) Effects of intensive exercise training on myocardial performance and coronary blood flow. J Appl Physiol 49:444–449

Bassey EJ, Fentem PH (1981) Work physiology. In: Edholm OG, Weiner JS (eds) The principles and practice of human physiology. Academic, London, pp 19–110

Bell RD, Rasmussen RL (1974) Exercise and the myocardial capillary fiber ratio during growth. Growth 38:237–244

Berkenboom GM, Abramowicz M, Vandermoten P, Degre SG (1986) Role of alpha-adrenergic coronary tone in exercise-induced angina pectoris. Am J Cardiol 57:195–198

Berne RM, Rubio R (1979) Coronary circulation. In: Berne RM, Sperelakis N, Geiger SR (eds) Handbook of physiology. American Physiological Society, Washington, DC, pp 873–952

Bersohn MM, Scheuer J (1977) Effects of physical training on end-diastolic volume and myocardial performance of isolated rat hearts. Circ Res 40:510–516

Bishop N, Hart G, Elamin MS, Silverton NP, Boyle R, Stoker JB, Smith DR, Mary DASG, Linden RJ (1986a) Assessment of the effect of nifedipine on myocardial ischaemia by using the ST segment/heart rate slope. Clin Sci 70:601–609

Bishop N, Linden RJ, Mary DASG, Stoker JB (1986b) The effect of vasodilator drugs on myocardial ischaemia in stable angina, using the maximal ST/HR slope. Clin Sci 70 [Suppl 13]:8P

Bishop N, Mackintosh AF, Stoker JB, Mary DASG, Linden RJ (1986c) The effect of indoramin on exercise performance in mild hypertension. J Cardiovasc Pharmacol 8 [Suppl 2]:S30–S36

Bishop N, Boyle RM, Stoker JB, Mary DASG (1987) The ST segment/heart rate relationship as an index of myocardial ischaemia. Int J Cardiol 14:281–293

Bjork L, O'Keefe A (1976) Estimation of coronary artery stenosis. Acta Radiol Diagn 17:777–780

Blomqvist CG, Saltin B (1983) Cardiovascular adaptations to physical training. Annu Rev Physiol 45:169–189

Bloor CM, Leon AS (1970) Interaction of age and exercise on the heart and its blood supply. Lab Invest 22:160–165

Bloor CM, Pasyk S, Leon AS (1970) Interaction of age and exercise on organ and cellular development. Am J Pathol 58:185–199

Bloor CM, White FC, Sanders TM (1984) Effects of exercise on collateral development in myocardial ischemia in pigs. J Appl Physiol 56:656–665

Bove AA, Hultgren PB, Ritzer TF, Carey RA (1979) Myocardial blood flow and hemodynamic responses to exercise training in dogs. J Appl Physiol 46:571–578

Breisch EA, White FC, Nimmo LA, McKirnan MD, Bloor CM (1986) Exercise-induced cardiac hypertrophy: a correlation of blood flow and microvasculature. J Appl Physiol 60:1259–1267

Burt JJ, Jackson R (1965) The effects of physical exercise on the coronary collateral circulation of dogs. J Sports Med Phys Fitness 5:203–208

Buttrick PM, Levite HA, Schaible TF, Ciambrone G, Scheuer J (1985a) Early increases in coronary vascular reserve in exercised rats are independent of cardiac hypertrophy. J Appl Physiol 59:1861–1865

Buttrick PM, Schaible TF, Scheuer J (1985 b) Combined effects of hypertension and conditioning on coronary vascular reserve in rats. J Appl Physiol 60:275–279

Capasso JM, Remily RM, Sonnenblick EH (1982) Alterations in mechanical properties of rat papillary muscle during maturation. Am J Physiol 242:H359–H364

Carew TS, Covell JW (1978) Left ventricular function in exercise-induced hypertrophy in dogs. Am J Cardiol 42:82–88

Carey RA, Tipton CM, Lund DR (1976) Influence of training on myocardial responses of rats subjected to conditions of ischaemia and hypoxia. Cardiovasc Res 10:359–367

Carey RA, Santamore WP, Michele JJ, Bove AA (1983) Effects of endurance training on coronary resistance in dogs. Med Sci Sports Exerc 15:355–359

Clausen JP (1976) Circulatory adjustments to dynamic exercise and effect of physical training in normal subjects and in patients with coronary artery disease. Prog Cardiovasc Dis 18:459–495

Clausen JP (1977) Effect of physical training on cardiovascular adjustments to exercise in man. Physiol Rev 57:779–815

Codini MA, Yipintsoi T, Scheuer J (1977) Cardiac responses to moderate training in rats. J Appl Physiol 42:262–266

Cohen MV, Yipintsoi T, Malhotra A, Penpargkul S, Scheuer J (1978) Effect of exercise on collateral development in dogs with normal coronary arteries. J Appl Physiol 45:797–805

Convertino V, Hung J, Goldwater D, de Busk RF (1982) Cardiovascular responses to exercise in middle-aged men after 10 days of bedrest. Circulation 65:134–140

Convertino VA, Keil LC, Greenleaf JE (1983) Plasma volume, renin, and vasopressin responses to graded exercise after training. J Appl Physiol 54:508–514

Cox ML, Bennett JB, Dudley GA (1986) Exercise training-induced alterations of cardiac morphology. J Appl Physiol 61:926–931

Coyle EF, Hemmert MK, Coggan AR (1986) Effects of detraining on cardiovascular responses to exercise: role of blood volume. J Appl Physiol 60:95–99

Crews J, Aldinger EE (1967) Effect of chronic exercise on myocardial function. Am Heart J 74:536–542

Cutilletta AF, Edmiston K, Dowell RT (1979) Effect of a mild exercise program on myocardial function and the development of hypertrophy. J Appl Physiol 46:354–360

Daly MDeB, Scott MJ (1962) An analysis of the primary cardiovascular reflex effects of stimulation of the carotid body chemoreceptors in the dog. J Physiol (Lond) 162:555–573

De Maria AN, Neumann A, Lee G, Fowler W, Mason DT (1978) Alterations in ventricular mass and performance induced by exercise training in man evaluated by echocardiography. Circulation 57:237–244

Denenberg DL (1972) The effects of exercise on the coronary collateral circulation. J Sports Med Phys Fitness 12:76–81

Detry J-M, Bruce AB (1971) Effects of physical training on exertional S-T-segment depression in coronary heart disease. Circulation 44:390–396

Dowell RT, Cutilletta AF, Rudnik MA, Sodt PC (1976) Heart functional responses to pressure overload in exercised and sedentary rats. Am J Physiol 230:199–204

Dowell RT, Stone HL, Sordahl LA, Asimakis GK (1977) Contractile function and myofibrillar ATPase activity in the exercise-trained dog heart. J Appl Physiol 43:977–982

Eckstein RW (1957) Effect of exercise and coronary artery narrowing on coronary collateral circulation. Circ Res 5:230–235

Ehsani AA, Hagberg JM, Hickson RC (1978) Rapid changes in left ventricular dimensions and mass in response to physical conditioning and deconditioning. Am J Cardiol 42:52–56

Ehsani AA, Heath GW, Hagberg JM, Sobel BE, Holloszy JO (1981) Effects of 12 months of intense exercise training on ischemic ST-segment depression in patients with coronary artery disease. Circulation 64:1116–1124

Ehsani AA, Heath GW, Martin WH, Hagberg JM, Holloszy JO (1984) Effects of intense exercise training on plasma catecholamines in coronary patients. J Appl Physiol 57:154–159

Ehsani AA, Biello DR, Schultz J, Sobel BE, Holloszy JO (1986) Improvement of left ventricular contractile function by exercise training in patients with coronary artery disease. Circulation 74:350–358

Ekblom B, Huot R (1972) Response to submaximal and maximal exercise at different levels of carboxyhemoglobin. Acta Physiol Scand 86:474–482

Ekelund L-G (1969) Exercise, including weightlessness. Annu Rev Physiol 31:85–116

Elamin MS, Winter C, Kardash MM, Silverton PN, Whitaker W, Smith DR, Mary DASG, Linden RJ (1983) Assessment of the effect of moderate exercise training on coronary heart disease using exercise ST segment/heart rate slope. Clin Sci 64:45P

Factor SM, Okun EM, Minase T, Kirk ES (1982) The microcirculation of the human heart: end-capillary loops with discrete perfusion fields. Circulation 66:1241–1248

Feigl EO (1983) Coronary physiology. Physiol Rev 63:1–205

Ferguson RJ, Petitclerc R, Choquette G, Chaniotis L, Gauthier P, Huot R, Allard C, Jankowski L, Campeau L (1974) Effect of physical training on treadmill exercise capacity, collateral circulation and progression of coronary disease. Am J Cardiol 34:764–769

Ferguson RJ, Cote P, Gauthier P, Bourassa MG (1978) Changes in exercise coronary sinus blood flow with training in patients with angina pectoris. Circulation 58:41–47

Finkelhor RS, Newhouse KE, Vrobel TR, Miron SD, Bahler RC (1986) The ST segment/heart rate slope as a predictor of coronary artery disease: comparison with quantitative thallium imaging and conventional ST segment criteria. Am Heart J 112:296–304

Flameng W, Schwartz F, Schaper W (1979) Coronary collaterals in the canine heart: development and functional significance. Am Heart J 97:70–77

Folkow B, Neil E (1971) Circulation. Oxford University Press, London

Froelicher VF (1972) Animal studies of effect of chronic exercise on the heart and atherosclerosis. Am Heart J 84:496–506

Froelicher VF (1973) The hemodynamic effects of physical conditioning in healthy young and middle-aged individuals, and in coronary heart disease patients. In: Naughton J, Hellerstein HK (eds) Exercise testing and exercise training in coronary heart disease. Academic, London, pp 63–77

Froelicher VF (1983) Exercise testing and training: clinical applications. J Am Coll Cardiol 1:114–125

Froelicher V, Battler A, McKirnan MD (1980) Physical activity and coronary heart disease. Cardiology 65:153–190

Fuller EO, Nutter DO (1981) Endurance training in the rat: II. Performance of isolated and intact heart. J Appl Physiol 51:941–947

Gattullo D, Linden RJ, Losano G, Mary DASG, Rosettani E, Soardo GP, Vacca G (1986) The effect of the coronary vascular wall of the changes in the coronary vascular resistance during a sudden reduction and recovery of the aortic blood pressure. Q J Exp Physiol 71:657–674

Gibson DG (1984) Study of left ventricular function in man by echocardiography. In: Techniques in the life sciences, vol. P3/II. Elsevier Scientific Publishers Ireland Ltd. Linden RJ (ed). Cardiovasc Physiol P319:1–37

Giusti R, Bersohn MM, Malhotra A, Scheuer J (1978) Cardiac function and actomyosin ATPase activity in hearts of conditioned and deconditioned rats. J Appl Physiol 44:171–174

Gleeson TT, Mullin WJ, Baldwin KM (1983) Cardiovascular responses to treadmill exercise in rats: effects of training. J Appl Physiol 54:789–793

Greenberg MA, Arbeit S, Rubin IL (1979) The role of physical training in patients with coronary artery disease. Am Heart J 97:527–534

Gregg DE, Fisher LC (1963) Blood supply to the heart. In: Hamilton WF, Dow P (eds) Circulation: handbook of physiology. American Physiological Society, Washington, DC, pp 1517–1584

Grimm AF, Kubota R, Whitehorn WV (1963) Properties of myocardium in cardiomegaly. Circ Res 12:118–124

Gwirtz PA, Stone HL (1984) Coronary vascular response to adrenergic stimulation in exercise-conditioned dogs. J Appl Physiol 57:315–320

Hagberg JM, Ehsani AA, Holloszy JO (1983) Effect of 12 months of intense exercise training on stroke volume in patients with coronary artery disease. Circulation 67:1194–1199

Hakkila J (1955) Studies on the myocardial capillary concentration in cardiac hypertrophy due to training: an experimental study with guinea pigs. Ann Med Exp Biol Fenn 32 [Suppl 10]:1–82

Halpern MH, May MM (1958) Phylogenetic study of the extracardiac arteries to the heart. Am J Anat 102:469–480

Hansford RG (1978) Lipid oxidation by heart mitochondria from young adult and senescent rats. Biochem J 170:285–295

Harpur RP (1980) The rat as a model for physical fitness studies. Comp Biochem Physiol 66:553–574

Harrison MH (1985) Effects of thermal stress and exercise on blood volume in humans. Physiol Rev 65:149–209

Haslam RW, Stull GA (1974) Duration and frequency of training as determinant of coronary tree capacity in the rat. Res Q Am Assoc Health Phys Ed 45:178–184

Hearse DJ (1979) Cellular damage during myocardial ischaemia: metabolic changes leading to enzyme leakage. In: Hearse DJ, de Leiris J (eds) Enzymes in cardiology. Wiley, New York, pp 1–19

Heaton WH, Marr KC, Capurro NL, Goldstein RE, Epstein SE (1978) Beneficial effect of physical training on blood flow to myocardium perfused by chronic collaterals in the exercising dog. Circulation 57:575–581

Hellerstein HK (1969) Relation of exercise to acute myocardial infarction. Therapeutic, restorative, preventive, and etiological aspects. Circulation 39, 40 [Suppl IV]:124–129

Hellerstein HK, Burlando A, Hirsch EZ, Plotkin FH, Feil GH, Winkler O, Marik S, Margolis N (1965) Active physical reconditioning of coronary patients (P). Circulation 31, 32 [Suppl II]:110–111

Hermansen L, Ekblom B, Saltin B (1970) Cardiac output during submaximal and maximal treadmill and bicycle exercise. J Appl Physiol 29:82–86

Hespel P, Lijnen P, Vanhees L, Fagard R, Amery A (1986) β-adrenoceptors and the regulation of blood pressure and plasma renin during exercise. J Appl Physiol 60:108–113

Hickson RC, Hammonds GT, Holloszy JO (1979) Development and regression of exercise-induced cardiac hypertrophy in rats. Am J Physiol 236:H268–H272

Hickson RC, Galassi TM, Dougherty KA (1983) Repeated development and regression of exercise-induced cardiac hypertrophy in rats. J Appl Physiol 54:794–797

Hoffman JIE (1981) Why is myocardial ischaemia so commonly subendocardial? Clin Sci 61:657–662

Hoffman JIF, Payne BD, Heymann MA, Rudolph AM (1983) The use of microspheres to measure blood flow. In: Techniques in the life sciences, vol. P3/1. Elsevier Scientific Publishers Ltd. Linden RJ (ed). Cardiovasc Physiol P304:1–36

Holloszy JO, Coyle EF (1984) Adaptations of skeletal muscle to endurance exercise and their metabolic consequences. J Appl Physiol 56:831–838

Holmgren A (1967) Cardiorespiratory determinants of cardiovascular fitness. Can Med Ass J 96:697–705

Hudlicka O (1982) Growth of capillaries in skeletal and cardiac muscle. Circ Res 50:451–461

Hung J, Goldwater D, Convertino VA, McKillop JH, Goris ML, de Busk RF (1983) Mechanisms for decreased exercise capacity after bed rest in normal middle-aged men. Am J Cardiol 51:344–348

Kalpinsky E, Hood WB, McCarthy B, McCombs L, Lown B (1968) Effects of physical training in dogs with coronary artery ligation. Circulation 37:556–565

Kappagoda CT, Linden RJ, Newell JP (1979) Effect of Canadian Air Force training programme on a submaximal exercise test. Q J Exp Physiol 64:185–204

Kennedy CC, Spiekerman RE, Lindsay MI, Mankin MT, Frye RL, McCallister BD (1976) One-year graduated exercise program for men with angina pectoris: evolution by physiologic studies and coronary arteriography. Mayo Clin Proc 51:231–236

Kirchheim HR (1976) Systemic arterial baroreceptor reflexes. Physiol Rev 56:100–176

Kligfield P, Okin PM, Ameisen O, Wallis J, Borer JS (1985) Correlation of the exercise ST/HR slope with anatomic and radionuclide cineangiographic findings in stable angina pectoris. Am J Cardiol 56:418–421

Kligfield P, Okin PM, Ameisen O, Borer JS (1986) Evaluation of coronary artery disease by an improved method of exercise electrocardiography: the ST segment/heart rate slope. Am Heart J 112:589–598

Kloner RA, Kloner JA (1981) The effect of exercise on healing of myocardial infarction. Circulation 64 [Suppl IV]:99

Koerner JE, Terjung RL (1982) Effect of physical training on coronary collateral circulation of the rat. J Appl Physiol 52:376–387

Kramer K, Lockner W, Wetterer E (1963) Methods of measuring blood flow. In: Hamilton WF, Dow P (eds) Circulation: handbook of physiology. American Physiological Society, Washington, DC, pp 1277–1324

Kramsch DM, Aspen AJ, Abramowitz BM, Kreimendahl T, Hood WB (1981) Reduction of coronary atherosclerosis by moderate conditioning exercise in monkeys on an atherogenic diet. N Engl J Med 305:1483–1489

Lakatta EG, Yin FCP (1982) Myocardial aging: functional alterations and related cellular mechanics. Am J Physiol 242:H927–H941

Laughlin MH (1985) Effects of exercise training on coronary transport capacity. J Appl Physiol 58:468–476

Laughlin MH, Diana JN (1975) Myocardial transcapillary exchange in the hypertrophied heart of the dog. Am J Physiol 229:838–846

Laughlin MH, Diana JN, Tipton CM (1978) Effects of exercise training on coronary reactive hyperemia and blood flow in the dog. J Appl Physiol 45:604–610

Lee AP, Ice R, Blessey R, Sanmarco ME (1979) Long-term effects of physical training on coronary patients with impaired ventricular function. Circulation 60:1519–1526

Lee JD, Tajimi T, Guth B, Seitelberger R, Miller M, Ross J (1986) Exercise-induced regional dysfunction with subcritical coronary stenosis. Circulation 73:596–605

Leon AS, Blackburn H (1977) The relationship of physical activity to coronary heart disease and life expectancy. Ann NY Acad Sci 301:561–578

Leon AS, Bloor CM (1968) Effects of exercise and its cessation on the heart and its blood supply. J Appl Physiol 24:485–490

Leon AS, Bloor CM (1976) The effect of complete and partial deconditioning on exercise-induced cardiovascular changes in the rat. Adv Cardiol 18:81–92

Leonard E, Hajdu S (1962) Action of electrolytes and drugs on the contractile mechanisms of the cardiac muscle cell. In: Hamilton WF, Dow P (eds) Circulation: handbook of physiology. American Physiological Society, Washington, DC, pp 151–197

Letac B, Cribier A, Desplanches JF (1977) A study of left ventricular function in coronary patients before and after physical training. Circulation 56:375–378

Lewis SF, Taylor WF, Graham RM, Pettinger WA, Schutte JE, Blomqvist CG (1983) Cardiovascular responses to exercise as functions of absolute and relative work load. J Appl Physiol 54:1314–1323

Li YX, Lincoln T, Mendelowitz D, Grossman W, Wei JY (1986) Age-related differences in effect of exercise training on cardiac muscle function in rats. Am J Physiol 251:H12–H18

Liang IYS, Stone HL (1982) Effect of exercise conditioning on coronary resistance. J Appl Physiol 53:631–636

Liang IYS, Stone HL (1983) Changes in diastolic coronary resistance during submaximal exercise in conditioned dogs. J Appl Physiol 54:1057–1062

Liang IYS, Hamra M, Stone HL (1984) Maximum coronary blood flow and minimum coronary resistance in exercise-trained dogs. J Appl Physiol 56:641–647

Linden RJ, Mary DASG (1982) Limitations and reliability of exercise electrocardiography tests in coronary heart disease. Cardiovasc Res 16:675–710

Linden RJ, Mary DASG (1983) The preparation and maintenance of anaesthetized animals for the study of cardiovascular function. In: Techniques in the life sciences, vol. P3/1. Cardiovasc Physiol P301:1–22

Ljungqvist A, Unge G (1972) The finer intramyocardial vasculature in various forms of experimental cardiac hypertrophy. Acta Pathol Microbiol Scand 80A:329–340

Ljungqvist A, Unge G (1973) The proliferative activity of the myocardial tissue in various forms of experimental cardiac hypertrophy. Acta Pathol Microbiol Scand 81A:233–240

Ljungqvist A, Unge G (1977) Capillary proliferative activity in myocardium and skeletal muscle of exercised rats. J Appl Physiol 43:306–307

Ljungqvist A, Unge G, Carlsson (1976) The myocardial capillary vasculature in exercising animals with increased cardiac pressure load. Acta Pathol Microbiol Scand 84A:244–246

Loguens RP, Gomez-Dumm CLA (1967) Fine structure of myocardial mitochondria in rats after exercise for one-half to two hours. Circ Res 11:271–279

Luksic IY, Raffo JA, Mary DASG, Watson DA, Deverall PB, Linden RJ (1981) Use of exercise tests in assessment of the functional result of aortocoronary bypass surgery. Thorax 36:428–434

Lynch P, Crawford IC (1983) Scintigraphic evidence of improvement in myocardial perfusion associated with improvement in physical fitness in a patient with angina. J R Army Med Corps 129:54–58

MacIntosh AM, Baldwin KM, Herrick RE, Mullin WM (1985) Effects of training on biochemical and functional properties of rodent neonatal heart. J Appl Physiol 59:1440–1445

Malhotra MS, Gupta JS, Joseph NT (1973) Comparative evaluation of different training programmes on physical fitness. Ind J Physiol Pharmacol 17:356–363

Mandache E, Unge G, Ljungqvist A (1972) Myocardial blood capillary reaction in various forms of cardiac hypertrophy. An electron microscopical investigation in the rat. Virchows Arch [Cell Pathol] 11:97–110

Mandache E, Unge G, Appelgren L-E, Ljungqvist A (1973) The proliferative activity of the heart tissues in various forms of experimental cardiac hypertrophy studied by electron microscope autoradiography. Virchows Arch [Cell Pathol] 12:112–122

Marsland WP (1968) Heart rate response to submaximal exercise in the standardbred horse. J Appl Physiol 24:98–101

Mary DASG (1986) Editorial note: Physiological mechanisms of changes in left ventricular performance during exercise in six subjects: editorial note. Int J Cardiol 10:233–235

Mary DASG, Winter C, Linden RJ (1986) Type of exercise training and cardiorespiratory fitness using the heart rate/oxygen consumption relationship. Proceed XXX Int Union Physiol Sci 16:216

Maseri A (1975) Myocardial blood flow in acute ischaemia, and its measurement. In: Oliver MF (ed) Modern trends in cardiology. Butterworths, London, pp 115–153

Mazzeo RS, Brooks GA, Horvath SM (1984) Effects of age on metabolic responses to endurance training in rats. J Appl Physiol 57:1369–1374

McArdle WD (1967) Metabolic stress of endurance swimming in the laboratory rat. J Appl Physiol 22:50–54

McArdle WD, Montoye HJ (1966) Reliability of exhaustive swimming in the laboratory rat. J Appl Physiol 21:1431–1434

McElroy CL, Gissen SA, Fishbein MC (1978) Exercise-induced reduction in myocardial infarct size after coronary artery occlusion in the rat. Circulation 57:958–962

Mirvis DM, Gordey RL (1983) Electrocardiographic effects of myocardial ischemia induced by atrial pacing in dogs with coronary stenosis: I. Repolarization changes with progressive left circumflex coronary artery narrowing. J Am Coll Cardiol 1:1090–1098

Mirvis DM, Ramanathan KB, Wilson JL (1986) Regional blood flow correlates of ST segment depression in tachycardia-induced myocardial ischemia. Circulation 73:365–373

Mole PA (1978) Increased contractile potential of papillary muscles from exercise-trained rat hearts. Am J Physiol 234:H421–H425

Mortimer IL, Reed JW (1982) Prediction of maximal oxygen uptake from submaximal blood lactate cocentration. J Physiol 328:73P

Musch TI, Haidet GC, Ordway GA, Longhurst JC, Mitchell JH (1985) Dynamic exercise training in foxhounds. I. Oxygen consumption and hemodynamic responses. J Appl Physiol 59:183 – 189

Musch TI, Moore RL, Leather DJ, Bruno A, Zelis R (1986) Endurance training in rats with chronic heart failure induced by myocardial infarction. Circulation 74:431 – 441

Neill WA, Oxendine JM (1979) Exercise can promote coronary collateral development without improving perfusion of ischemic myocardium. Circulation 60:1513 – 1519

Newell JP (1982) The physical rehabilitation of patients after cardiac surgery. Ph. D. Thesis, University of Leeds

Newell JP, Kappagoda CT, Stoker JB, Deverall PB, Watson DA, Linden RJ (1980) Physical training after heart valve replacement. Br Heart J 44:638 – 649

Newman PE (1981) The coronary collateral circulation: determinants and functional significance in ischemic heart disease. Am Heart J 102:431 – 445

Nolewajka AJ, Kostuk WJ, Rechnitzer PA, Cunningham DA (1979) Exercise and human collateralization: an angiographic and scintigraphic assessment. Circulation 60:114 – 121

Nutter DO, Fuller EO (1977) The role of isolated cardiac muscle preparations in the study of training effects on the heart. Med Sci Sports Exerc 9:239 – 245

Nutter DO, Priest RE, Fuller EO (1981) Endurance training in the rat: I. Myocardial mechanics and biochemistry. J Appl Physiol 51:934 – 940

O'Brien DW (1981) The effect of prolonged physical training and high fat diet on heart size and body weight in rats. Can J Physiol Pharmacol 59:268 – 272

Okin PM, Kligfield P, Ameisen O, Goldberg HL, Borer JS (1985) Improved accuracy of the exercise electrocardiogram: identification of three-vessel coronary disease in stable angina pectoris by analysis of peak rate related changes in ST segments. Am J Cardiol 55:271 – 276

Okin PM, Ameisen O, Kligfield P (1986) A modified treadmill exercise protocol for computer-assisted analysis of the ST segment/heart rate slope: methods and reproducibility. J Electrocardiology 19:311 – 318

Okun EM, Factor SM, Kirk ES (1979) End-capillary loops in the heart: an explanation for discrete myocardial infarctions without border zones. Science 206:565 – 567

Ordway GA, Floyd DL, Longhurst JC, Mitchell JH (1984) Oxygen uptake and hemodynamic responses during graded exercise in the dog. J Appl Physiol 57:601 – 607

Oscai LB, Mole PA, Brei B, Holloszy JO (1971 a) Cardiac growth and respiratory enzyme levels in male rats subjected to a running program. Am J Physiol 220:1238 – 1241

Oscai LB, Mole PA, Holloszy JO (1971 b) Effects of exercise on cardiac weight and mitochondria in male and female rats. Am J Physiol 220:1944 – 1948

Parizkova J, Wachtlova M, Soukupova M (1972) The impact of different motor activity on body composition, density of capillaries and fibres in the heart and soleus muscles and cell's migration in vitro in male rats. Int Z Angew Physiol 30:207 – 216

Parsons D, Musch TI, Moore RL, Haidet GC, Ordway GA (1985) Dynamic exercise in foxhounds: II. Analysis of skeletal muscle. J Appl Physiol 59:190 – 197

Penpargkul S, Scheuer J (1970) The effect of physical training upon the mechanical and metabolic performance of the rate heart. J Clin Invest 49:1859 – 1968

Pollock ML (1973) The quantification of endurance training programs. In: Wilmore JH (ed) Exercise and sport science reviews. Academic, New York, pp 155 – 188

Raffo JA, Luksic IY, Kappagoda CT, Mary DASG, Whitaker W, Linden RJ (1980) Effects of physical training on myocardial ischaemia in patients with coronary artery disease. Br Heart J 43:262 – 269

Rakusan K, Ostadal B, Wachtlova M (1971) The influence of muscular work on the capillary density in the heart and skeletal muscle of pigeon (*Columba dom.*). Can J Physiol Pharmacol 49:168 – 170

Rauramaa R, Salonen JT, Kukkonen-Harjula K, Seppanen K, Seppala E, Vapaatalo H, Huttunen JK (1984) Effects of mild physical exercise on serum lipoproteins and metabolites of arachidonic acid: a controlled randomised trial in middle aged men. Br Med J 288:603 – 606

Redwood DR, Rosing DR, Epstein SE (1972) Circulatory and symptomatic effects of physical training in patients with coronary-artery disease and angina pectoris. N Engl J Med 286:959–965

Reimer KA, Ideker RE, Jennings RB (1981) Effect of coronary occlusion site on ischaemic bed size and collateral blood flow in dogs. Cardiovasc Res 15:668–674

Restorff W, Holtz J, Bassenge E (1977) Exercise induced augmentation of myocardial oxygen extraction in spite of normal dilatory capacity in dogs. Pflugers Arch 372:181–185

Riedhammer HH, Rafflenbeul W, Weihe WH, Krayenbuhl HP (1976) Left ventricular contractile function in trained dogs with cardial hypertrophy. Basic Res Cardiol 71:297–308

Rigotti NA, Thomas GS, Leaf A (1983) Exercise and coronary heart disease. Annu Rev Med 34:391–412

Ritzer TF, Bove AA, Carey RA (1980) Left ventricular performance characteristics in trained and sedentary dogs. J Appl Physiol 48:130–138

Roskamm H (1967) Optimum patterns of exercise for healthy adults. Can Med Assoc J 96:895–900

Rowell LB (1974) Human cardiovascular adjustments to exercise and thermal stress. Physiol Rev 54:75–159

Saltin B, Blomqvist G, Mitchell JE, Johnson RL, Wildenthal K, Chapman CB (1968) Response to exercise after bed rest and after training: a longitudinal study of adaptive changes in oxygen transport and body composition. Circulation 37, 38 [Suppl VII]:1–78

Sanders M, White FC, Peterson TM, Bloor CM (1978) Effects of endurance exercise on coronary collateral blood flow in miniature swine. Am J Physiol 234:H614–H619

Sarnoff SJ, Mitchell JH, Gilmore JP, Remensnyder JP (1960) Homeometric autoregulation in the heart. Circ Res 8:1077–1091

Schaible TF, Scheuer J (1979) Effects of physical training by running or swimming on ventricular performance of rat hearts. J Appl Physiol 46:854–860

Schaible TF, Scheuer J (1981) Cardiac function in hypertrophied hearts from chronically exercised female rats. J Appl Physiol 50:1140–1145

Schaible TF, Scheuer J (1985) Cardiac adaptations to chronic exercise. Prog Cardiovasc Dis 27:297–324

Schaible TF, Penpargkul S, Scheuer J (1981) Cardiac responses to exercise training in male and female rats. J Appl Physiol 50:112–117

Schaper W (1978) Experimental coronary artery occlusion: III. The determinants of collateral blood flow in acute coronary occlusion. Basic Res Cardiol 73:584–594

Schaper W (1982) Influence of physical exercise on coronary collateral blood flow in chronic experimental two-vessel occlusion. Circulation 65:905–912

Schaper W, Wusten B (1979) Collateral circulation. In: Schaper W (ed) The pathophysiology of myocardial perfusion. Elsevier/North-Holland, Amsterdam, pp 415–470

Schaper W, Flameng W, Brabander MD (1972) Comparative aspects of coronary collateral circulation. Adv Exp Med Biol 22:267–276

Scheel KW, Ingram LA, Wilson JL (1981) Effects of exercise on the coronary and collateral vasculature of beagles with and without coronary occlusion. Circ Res 48:523–530

Scheuer J (1982) Effects of physical training on myocardial vascularity and perfusion. Circulation 66:491–495

Scheuer J, Tipton CM (1977) Cardiovascular adaptations to physical training. Annu Rev Physiol 39:221–225

Scheuer J, Penpargkul S, Bhan AK (1974) Experimental observations on the effects of physical training upon intrinsic cardiac physiology and biochemistry. Am J Cardiol 33:744–751

Shephard RJ (1978) Methods for the measurement of physical fitness, working capacity and activity patterns. In: Shephard RJ (ed) Human physiological work capacity. Cambridge University Press, Cambridge, pp 22–46

Siegel W, Blomqvist G, Mitchell JH (1970) Effects of a quantitated physical training program on middle-aged sedentary men. Circulation 41:19–29

Sinning WE (1975) Factors of fitness. In: Wilson PK (ed) Adult fitness and cardiac rehabilitation. University Park, Baltimore, pp 29–42

Sim DN, Neill WA (1974) Investigation of the physiological basis for increased exercise threshold for angina pectoris after physical conditioning. J Clin Invest 54:763–770

Sonnenblick EH (1962) Force-velocity relations in mammalian heart muscle. Am J Physiol 202:931–939

Sonnenblick EH, Parmley WW, Urschel CW (1969) The contractile state of the heart as expressed by force-velocity relations. Am J Cardiol 23:488–503

Spear KL, Koerner JE, Terjung RL (1978) Coronary blood flow in physically trained rats. Cardiovasc Res 12:135–143

Spiro SG, Juniper E, Bowman P, Edwards RHT (1974) An increasing work rate test for assessing the physiological strain of submaximal exercise. Clin Sci 46:191–206

Spurgeon HA, Steinbach MF, Lakatta EG (1983) Chronic exercise prevents characteristic age-related changes in rat cardiac contraction. Am J Physiol 244:H513–H518

Starnes JW, Beyer RE, Edington DW (1983) Myocardial adaptations to endurance exercise in aged rats. Am J Physiol 245:H560–H566

Stevenson JAF, Feleki V, Rechnitzer P, Beaton JR (1964) Effect of exercise on coronary tree size in the rat. Circ Res 15:265–269

Stone L (1977) Cardiac function and exercise training in conscious dogs. J Appl Physiol 42:824–832

Stone HL (1980a) The heart and exercise training. In: Bourne GH (ed) Hearts and heart-like organs, vol 2. Academic, London, pp 389–418

Stone HL (1980b) Coronary flow, myocardial oxygen consumption, and exercise training in dogs. J Appl Physiol 49:759–768

Templeton GH, Platt MR, Willerson JT, Weisfeldt ML (1979) Influence of aging on left ventricular hemodynamics and stiffness in beagles. Circ Res 44:189–194

Tepperman J, Pearlman D (1961) Effects of exercise and anaemica on coronary arteries of small animals as revealed by the corrosion-cast technique. Circ Res 9:576–584

Thomas BT, Millar AT (1958) Adaptation to forced exercise in rats. Am J Physiol 193:350–354

Tibbits G, Koziol BJ, Roberts NK, Baldwin KM, Barnard RJ (1978) Adaptation of the rat myocardium to endurance training. J Appl Physiol 44:85–89

Tibbits GF, Barnard RJ, Baldwin KM, Cugalj N, Roberts NK (1981) Influence of exercise on excitation-contraction coupling in rat myocardium. Am J Physiol 240:H472–H480

Tipton CM (1965) Training and bradycardia in rats. Am J Physiol 209:1089–1094

Tipton CM, Carey RA, Eastin WC, Erickson HH (1974) A submaximal test for dogs: evaluation of effects of training, detraining, and cage confinement. J Appl Physiol 37:271–275

Tomanek RJ (1970) Effects of age and exercise on the extent of the myocardial bed. Anat Rec 167:55–62

Tomanek RJ, Banister EW (1972) Myocardial ultrastructure after acute exercise stress with special reference to transverse tubules and intercalated discs. Cardiovasc Res 6:671–679

Unge G, Carlsson S, Ljungqvist A, Tornling G, Adolfsson J (1979) The proliferative activity of myocardial capillary wall cells in variously aged swimming-exercised rats. Acta Pathol Microbiol Scand 87A:15–17

Vatner SF, Murray PA (1982) Reflex control of coronary arteries. In: Kalsner S (ed) The coronary artery. Croom-Helm, London, pp 216–238

Wackers TJTh (1984) Radionuclide techniques for assessment of cardiac function in man. In: Techniques in the life sciences, vol. P3/II. Elsevier Scientific Publishers Ireland Ltd Linden RJ (ed). Cardiovasc Physiol P320:1–26

Wallace AG, Rerych SK, Jones RH, Goodrich JK (1978) Effects of exercise training on ventricular function in coronary disease. Circulation 57, 58 [Suppl II]:197

Weller JJ, El-Gamal FM, Parker L, Reed JW, Bridges NG, Chinn DJ, Cotes JE (1985) Estimating the capacity for exercise of shipyard workers. Clin Sci 68 [Suppl 11]:45P

Wexler BC, Greenberg BP (1974) Effect of exercise on myocardial infarction in young vs. old male rats: electrocardiograph changes. Am Heart J 88:343–350

Williams JF, Potter RD (1976) Effect of exercise conditioning on the intrinsic contractile state of cat myocardium. Circ Res 39:425–428

Winkler B (1984) Measurement of coronary blood flow. In: Techniques in life sciences, Elsevier
    Scientific Publishers Ireland Ltd. Linden RJ (ed). Vol P3/II. Cardiovasc Physiol P316:1−36
Winter C, Kardash MM, Whitaker W, Mary DASG, Linden RJ (1984) The effects of long-term-
    in physical training in patients with coronary heart disease. Int J Cardiol 5:675−685
Woodson RD, Willis RE, Lenfant C (1978) Effect of acute and established anemia on O_2 trans-
    port at rest, submaximal and maximal work. J Appl Physiol 44:36−43
Wyatt HL (1982) Physical conditioning and the coronary-artery vasculature. In: Kalsner S (ed)
    The coronary artery. Croom-Helm, London, pp 365−388
Wyatt HL, Mitchell JH (1974) Influence of physical training on the heart of dogs. Circ Res
    35:883−889
Wyatt HL, Mitchell J (1978) Influences of physical conditioning and deconditioning on cor-
    onary vasculature of dogs. J Appl Physiol 45:619−625
Wyatt HL, Chuck L, Rabinowitz B, Tyberg JV, Parmley WW (1978) Enhanced cardiac response
    to catecholamines in physically trained cats. Am J Physiol 234:H608−H613
Wydenham CH (1967) Submaximal tests for estimating maximum oxygen intake. Can Med Ass
    J 96:736−745
Yipintsoi T, Rosenkrantz J, Codini MA, Scheuer J (1980) Myocardial blood flow responses to
    acute hypoxia and volume loading in physically trained rats. Cardiovasc Res 14:50−57

Rev. Physiol. Biochem. Pharmacol., Vol. 109
© by Springer-Verlag 1987

The Physiological Function of Nerve Growth Factor in the Central Nervous System: Comparison With the Periphery

HANS THOENEN, CHRISTINE BANDTLOW, and ROLF HEUMANN

Contents

Department of Neurochemistry; Max-Planck-Institute for Psychiatry, D-8033 Planegg-Martinsried, FRG
The references cited in this monograph cover the published literature up until October 1, 1986. Any work quoted after this date refers to articles originating from our own or related labs.

1 Introduction

Of all isolated neurotrophic molecules Nerve Growth Factor (NGF) is the only one with a fully established physiological function, at least in the peripheral
nervous system (see Thoenen and Edgar 1985; Thoenen et al. 1987b). This
is largely due to the fact that it is present in very large quantities in exocrine
glands (e.g. submandibular gland of the male mouse and accessory genital
organs of various species). The physiological function of NGF in these exocrine glands is unknown. In any case it is irrelevant for the development and
maintenance of specific properties of NGF responsive neurons (see Greene
and Shooter 1980; Thoenen and Barde 1980; Thoenen et al. 1985). These rich
sources allowed the purification of NGF at an early stage of NGF research.
This purification was an essential prerequisite for the production of anti-NGF
antibodies (see Levi-Montalcini 1966; Levi-Montalcini and Angeletti 1968),
the determination of NGF's amino acid sequence (Hogue-Angeletti and
Bradshaw 1971) and allowed the molecular cloning and establishment of the
genomic organization of NGF (Scott et al. 1983; Ullrich et al. 1983). Over the
last decades NGF research was predominantly focussed on the peripheral
target neurons of NGF, the sympathetic and the neural crest-derived sensory
neurons (see Levi-Montalcini and Angeletti 1968; Greene and Shooter 1980;
Thoenen and Barde 1980). Increasing attention has been diverted towards the
central nervous system, however, as it became apparent that the cholinergic
neurons of the basal forebrain nuclei are also targets for NGF (Schwab et al.
1979; Gnahn et al. 1983; Seiler and Schwab 1984). The possible links between
NGF and the pathophysiology and potential therapy of Alzheimer's disease
have thus moved into the range of realistic consideration (see Hefti 1983; Hefti and Weiner 1986).

In the following we will first give a brief survey of the most essential aspects
of the present state of knowledge of the functions of NGF in the peripheral
nervous system. Thenceforward, we will concentrate on those aspects which
are important for the evaluation of the physiological function of NGF in the
central nervous system. We will then present a review of the current state of
knowledge of the functions of NGF in the central nervous system and finally,
we will discuss these functions in the context of the pathophysiology and
potential therapy of Alzheimer's disease.

2 Survey of the Physiological Functions of NGF in the Peripheral Nervous System

2.1 Structure-Function Relationship of NGF Molecules from Different Species

For many years NGF research has been primarily based on NGF purified from the male mouse salivary gland and the antibodies produced against it. At a relatively early stage of NGF research it became apparent that injection of anti-mouse NGF antibodies into chick embryos did not result in the same extensive destruction of the sympathetic nervous system as observed after antibody injections into newborn mice and rats (see Levi-Montalcini and Angeletti 1968; Thoenen and Barde 1980). Since chick sympathetic and sensory neurons respond to mouse NGF in vivo and in vitro in a way similar to the corresponding mouse neurons, it was reasonable to conclude that the domain(s) of the NGF molecule responsible for its biological activity must have been preserved, whereas other domains had changed during evolution. This assumption was further substantiated when bovine NGF was purified from bovine seminal plasma (Harper et al. 1982), and a detailed and comprehensive comparison between the biological activity of pure mouse and bovine NGF became possible. These experiments demonstrated that the biological activity of mouse and bovine NGF were identical, although immunological crossreactivity was very limited (Harper et al. 1983). The molecular cloning of mouse, human, bovine, and chick NGF, together with amino acid sequence analysis of mouse NGF has allowed comparison of the conserved and unconserved domains of these molecules and their relationship to biological activity and antigenicity. The overall conservation of NGF during evolution is remarkably high. Of the 118 amino acids of mature mouse β-NGF, only 16 amino acids were changed in bovine, and 19 in chick NGF (Meier et al. 1986). As was expected from previous observations that the reduction of the three S-S bridges of mouse NGF led to a complete loss of biological activity (see Greene and Shooter 1980; Thoenen and Barde 1980), all the cysteine residues were strictly conserved. The apparent discrepancy between the overall high conservation of the amino acid sequence and the poor immunological crossreactivity is due to the fact that the amino acid changes between species are located in clusters. Hydropathy plots demonstrated that the changes are virtually exclusively located in the hydrophilic domains (Meier et al. 1986) expected to be potential antigenic determinants (see Hopp and Woods 1981). One single hydrophilic region has been shown to be strictly conserved in the NGF molecules of all species investigated so far (Meier et al. 1986). This conserved domain lends itself to future analysis by site-directed mutagenesis and by antibodies directed against synthetic peptides corresponding to this region.

2.2 Spectrum of Physiological Actions of NGF in the Peripheral Nervous System

NGF is essential for the regionally selective regulation of the survival of peripheral sympathetic and neural crest-derived sensory neurons during restricted periods of their development (Levi-Montalcini and Angeletti 1968; Thoenen and Barde 1980). Moreover, NGF is essential for the regulation of the expression of specific properties of these neurons during the period of neuronal differentiation. For instance, NGF regulates the synthesis of specific enzymes involved in the formation of catecholamines and neuron-specific peptides, such as substance P, somatostatin, and cholecystokinin (see Otten 1984; Thoenen et al. 1985). These regulatory actions of NGF on neuron-specific enzymes and peptides are not only of importance during the period of neuronal differentiation, but are also essential for the maintenance of neuron-specific properties in the adult.

2.3 NGF as a (Retrograde) Messenger Between Peripheral Target Organs and Innervating NGF-Responsive Neurons

NGF is taken up by sympathetic and sensory nerve terminals by a highly selective, saturable receptor-mediated mechanism followed by (rapid) retrograde transport in membrane-confined compartments to the perikarya (see Hendry 1980; Thoenen and Barde 1980; Schwab et al. 1982; Schwab and Thoenen 1983). It must be emphasized that under physiological conditions the receptors of NGF-responsive neurons are never saturated by endogenous NGF. This subsaturation of receptors is the prerequisite for all "pharmacological" effects of NGF resulting in an augmented survival (if administered during the period of natural cell death) and in the increase in neuron-specific enzymes and peptides which are also physiologically regulated by NGF (see Hendry 1980; Thoenen and Barde 1980; Schwab and Thoenen 1983). That these pharmacological effects also reflect physiologically relevant functions, was demonstrated with the observation that interference with retrograde axonal transport had the same effect as the neutralization of endogenous NGF by anti-NGF antibodies (see Hendry 1980; Thoenen and Barde 1980; Schwab and Thoenen 1983). During limited periods of the embryonic development, NGF deprivation (by administration of anti-NGF antibodies or interruption of the retrograde axonal transport) results in a degeneration of the corresponding sympathetic and sensory neurons (see Levi-Montalcini and Angeletti 1968; Johnson et al. 1980, 1986; Thoenen and Barde 1980). In fully differentiated neurons interference with the availability of NGF by the same procedures results in neuronal degeneration only under extreme conditions, i.e. very long-lasting NGF deprivation by autoimmunization (Gorin and Johnson 1979, 1980; Rich et al. 1984). An impairment of specialized functions is, however, consistently observed. For example, the administration of a single dose of antibodies (which in newborn animals results in an extensive destruction of the sympathetic nervous system) results in a marked but transient reduction of the synthesis of both tyrosine hydroxylase and dopamine-β-hydroxylase, key enzymes in the formation of the adrenergic transmitter noradrenaline (see Thoenen and Barde 1980; Schwab and Thoenen 1983).

Correspondingly, the administration of antibodies to adult animals also resulted in a reduction of neuron-specific peptides in the spinal sensory neurons (see Otten 1984).

Recently, this indirect evidence for a function of NGF as a retrograde messenger between peripheral effector organs and innervating NGF-responsive neurons was supplemented by more direct evidence. The development of a highly sensitive enzyme immunoassay for NGF (accurate to within 0.01 fmol of NGF/assay) made it possible to determine NGF levels in peripheral effector organs (Korsching and Thoenen 1983 a). Additionally, procedures were developed for the reliable quantification of the very rare mRNA^{NGF} (Heumann et al. 1984; Shelton and Reichardt 1984). These methods demonstrated a positive correlation between the density of sympathetic innervation and the levels of NGF (Korsching and Thoenen 1983 a) and its mRNA (Heumann et al. 1984; Shelton and Reichardt 1984). Moreover, the enzyme immunoassay for NGF also allowed a direct demonstration of the retrograde axonal transport of endogenous NGF (Korsching and Thoenen 1983 b) and thus provided evidence that the high levels of NGF in sympathetic ganglia result from accumulation of NGF by retrograde axonal transport rather than by local synthesis. This was concluded after the observation that the interruption of retrograde axonal transport by various procedures resulted in a rapid decay in NGF levels in sympathetic ganglia (Korsching and Thoenen 1985 b). The $t_{1/2}$ of the decay was 4–5 h. Accordingly, the levels of mRNA^{NGF} in sympathetic ganglia are at the limit of detectability, even though the NGF levels in these ganglia exceeded those of the most densely innervated peripheral organs, such as vas deferens, iris and heart atrium, by nearly an order of magnitude (Korsching and Thoenen 1983 a; Heumann et al. 1984).

2.4 Regulation of Synthesis and Cellular Localization of NGF in Peripheral Target Organs

As outlined above, the blockade of retrograde axonal transport by 6-hydroxydopamine (by the selective destruction of sympathetic nerve terminals) and colchicine (disassembly of microtubules) resulted in a rapid decrease of NGF in sympathetic ganglia ($t_{1/2}$ 4−5 h). This rapid decay in sympathetic ganglia was accompanied by a corresponding increase in NGF levels in peripheral target organs (Korsching and Thoenen 1985 b). These observations raised the question of whether the NGF increase in target tissues resulted exclusively from the elimination of the efficient removal of NGF by retrograde axonal transport or whether there was also an enhanced NGF synthesis in these tissues. This question was addressed in tissue culture experiments with the rat iris (Barth et al. 1984; Heumann and Thoenen 1986), a tissue which is innervated in vivo by sympathetic, parasympathetic and sensory neurons. After an initial lag period of 2 h, mRNA^{NGF} levels in cultured iris increased to a maximum of 35 times the base levels after 12 h. Thereafter levels declined to 2- to 3-fold the base after 48 h, but remained constant up to 72 h (Heumann and Thoenen 1986). This increase in mRNA^{NGF} levels was immediately followed by an enhanced NGF synthesis and release of NGF into the culture medium (Barth et al. 1984). The rapidly increasing mRNA^{NGF} levels up to 12 h are suggestive of a mechanism resulting from the release of molecules stored in nerve terminals. Nerve terminals are disconnected from their cell bodies by the culture procedure and therefore degenerate. Since it has been demonstrated that the process of nerve fiber degeneration starts earlier when the peripheral nerve stumps are shorter (Malmfors and Sachs 1965; Lubinska 1975), it can be assumed that the degeneration process starts virtually immediately after bringing the irides into culture. A large number of neuropeptides and aminergic transmitter substances has been investigated, but none of them is able to trigger an increase in NGF and mRNA^{NGF} comparable to that of the spontaneous increase in culture (Hellweg and Heumann unpublished results). As an alternative to the idea of release of constituents stored in nerve terminals, Shelton and Reichardt (1986 b) suggested that the augmented production of mRNA^{NGF} resulted from the tissue damage of the lesion procedure as such, since they found augmented mRNA^{NGF} levels only after retrobulbar general denervation (which seems to result in a damage to the iris beyond the interruption of the nerve supply), but not after selective transection of the sensory and sympathetic nerve fibers supplying the iris.

   The problem of the cellular localization of NGF synthesis has been resolved by developing highly selective and sensitive in-situ hybridization procedures with ^{35}S-labeled synthetic oligonucleotides, or cRNA probes produced in the SP6 system (for details see Bandtlow et al. 1987). These in situ-hybridization experiments demonstrated that the label was more or less equally distributed over all layers of the iris, including the surface epithelium. Reduced labelling, however, was consistently observed over the areas of the constrictor muscle. This is in agreement with the observation that the NGF and mRNA^{NGF} levels in the constrictor area are much lower than in the dilator area (Barth et al. 1984; Shelton and Reichardt 1986 b). If the in-situ hybridization experiments were performed in irides maintainted in culture for

12 h before cutting the cryostate sections, the overall distribution of the label was approximately the same, but the signal was much stronger; this is in agreement with augmented levels of mRNA^{NGF} that have been detected (Heumann and Thoenen 1986; Shelton and Reichardt 1986b). The resolution obtained in cryostate sections is insufficient to allow the association of label with individual cell types. Therefore, further experiments were performed with dissociated cultures of rat irides. There, all cells present in the iris, i.e. smooth muscle cells (identified by antibodies to desmin), fibroblasts (identified by antibodies to Thy-1), Schwann cells (identified by antibodies to O4-antigen) and even the small moiety of epithelial cells (identified by antibodies to keratin) showed NGF-specific labelling. These experiments demonstrate that differences between organs and the regional differences within organs (e.g. constrictor and dilator muscle of the iris) in the levels of expression of NGF cannot be associated with a single population of cells. It appears that the innervating neurons have an important regulatory function on the local synthesis of NGF. These in-situ hybridization experiments have also demonstrated that the suggestion of Rush and collaborators (Rush 1984; Finn et al. 1986) that NGF is predominantly, if not exclusively produced by Schwann cells (based on immunohistochemical data) is likely to be incorrect. The recent finding of Taniuchi et al. (1986) that adult Schwann cells re-express NGF receptors after denervation (in early stages of development all Schwann cells express NGF receptors (Zimmermann and Sutter 1983; Rohrer 1985)) suggests that NGF produced by both Schwann cells and other cells such as smooth muscle cells and fibroblasts could be bound by the NGF-receptors of the Schwann cells (or even be accumulated by internalization) thereby resulting in detectable staining with anti-NGF antibodies of Schwann cells only. This interpretation is further supported by the observation of Bandtlow et al. (1987) that binding of radioiodinated NGF to dissociated rat iris cells occurs exclusively on Schwann cells.

The regulatory function of neurons of the synthesis of NGF is not only confined to the peripheral target tissue of NGF-responsive neurons. It also seems to be exerted on the cells immediately surrounding the axons, that is, the Schwann cells and fibroblasts. This conclusion can be drawn from the results of recent experiments, which have demonstrated that NGF production in the adult rat sciatic nerve by Schwann cells and fibroblasts is very low under normal physiological conditions. The mRNA^{NGF} levels in the sciatic nerve (determined as fg of mRNA^{NGF} per gram wet weight) amount to only about 1/20 to 1/100 of those peripheral target tissues (Heumann et al. 1987). That the NGF produced by Schwann cells ensheathing axons is not essential for the neurons can be deduced from the observation that in newborn rats the selective destruction of sympathetic nerve terminals by 6-hydroxydopamine or the removal of the peripheral target tissues still leads to a degeneration of the corresponding sympathetic neurons. These degenerative changes, however, can be prevented by systemic administration of NGF (see Thoenen and Barde 1980; Schwab and Thoenen 1983). Moreover, if the production of NGF by Schwann cells did substantially contribute to the NGF supply of the NGF-responsive neurons in the sciatic nerve (which contains axons of both sensory and sympathetic neurons), a proximo-distal gradient of NGF would be expected. This, however, is not the case (Heumann et al. 1987). The low $mRNA^{NGF}$ levels in the sciatic nerve increase dramatically, however, after transection and the levels remain elevated for weeks in the stump as long as the regeneration of axons is prevented. Again, the very rapid transient increase of NGF levels in the regions immediately adjacent to the transection site favours the idea of a local effect associated with the lesion as suggested by Shelton and Reichardt (1986b) for the iris, whereas the permanent $mRNA^{NGF}$ elevation lasting for weeks speaks in favour of the normal repression of NGF synthesis by immediate contact with the axons. It should also be mentioned that in in-situ hybridization experiments with dissociated cells from the sciatic of both adult and newborn animals demonstrated a labelling of all the Schwann cells and fibroblasts (Bandtlow et al. 1987; Heumann et al. 1987). Thus, it seems that all Schwann cells and all fibroblasts produce NGF and not just those associated with the axons of NGF-responsive neurons, i.e. sympathetic and sensory neurons. Preliminary experiments have demonstrated that in newborn animals the $mRNA^{NGF}$ levels of the sciatic nerve are considerably higher than in adult animals, allowing the identification of NGF-producing cells by in-situ hybridization without transection (Bandtlow et al. 1987). These experiments demonstrated that $mRNA^{NGF}$ is localized around all of the axons and not only around a selective population, as would be expected if only those Schwann cells ensheathing sympathetic and sensory neurons expressed NGF. In this context it is worth mentioning that in newborn animals NGF receptors are also expressed by Schwann cells without transection (Heumann and Lindholm, unpublished observation). The possible function of these NGF-receptors is under investigation, in particular with respect to the regulation of the synthesis of NGF and molecules of the extracellular matrix such as laminin. The enhanced production of laminin could play an essential role in the process of neuronal regeneration, since it has been shown to be not only a good substrate for neuronal fiber outgrowth, but also to potentiate the neurotrophic action of NGF and of brain-derived neurotrophic factor (see Thoenen and Edgar 1985; Barde et al. 1987; Thoenen et al. 1987b). It is tempting to speculate that in newborn animals, where a larger production of NGF by Schwann cells would be expected, the NGF-responsive neurons cannot benefit to an appreciable extent from this production, for a large part of the NGF produced by the Schwann cells is also taken up and possibly stored and/or digested by them.

2.5 Mechanism of Action of Nerve Growth Factor on Peripheral Target Neurons

A broad spectrum of physiological actions of NGF has been described in much detail and yet, it is still not known by which mechanism(s) the interaction of NGF with its specific receptors is translated into the numerous short-, intermediate- and long-term effects (see Thoenen and Edgar 1985; Thoenen et al. 1985; Radeke et al. 1987). The kinetics of the interaction of NGF with its receptors have been studied extensively both in physiological target neurons and in NGF-responsive tumor cell lines, such as PC 12 phaeochromocytoma cells (see Sutter et al. 1984). The receptors on PC 12 cells have been identified by crosslinking ^{125}I-labelled NGF (Hosang and Shooter 1985) and purified by affinity chromatography (Puma et al. 1983). Moreover, monoclonal antibodies against PC 12 cell NGF receptors have been developed (Chandler et al. 1984), which also allowed for its cloning (Radeke et al. 1987). The molecular cloning of the NGF receptor has provided evidence that this receptor belongs to a new class of receptors (Radeke et al. 1987). This new class lacks the protein kinase consensus sequence of other "growth factors", which, in contrast to NGF, act as mitogens on their target cells (see Thoenen et al. 1985). The elucidation of the structure of the NGF receptor is an essential prerequisite for a more detailed analysis of the signal transduction mechanism of NGF. Previous experiments have demonstrated that a second messenger mechanism for NGF must exist (Heumann et al. 1981; Schwab et al. 1982). However, information on the nature of the second messenger(s) from these studies was exclusively negative; cAMP, calcium influx and the sodium/potassium activated ATPase could all be exluded (for discussion, see Thoenen et al. 1985). Recently, experiments with PC 12 cells have provided convincing evidence that a ras-like protein might be involved in the NGF-mediated signal transduction (Bar-Sagi and Feramisco 1985; Noda et al. 1985; Hagag et al. 1986). It will be of particular interest to see whether the observations made in phaeochromocytoma tumor cells can also be verified in physiological target cells of NGF, and to investigate how the NGF receptor is coupled to a ras-like protein.

Another promising approach to the elucidation of the mechanism of action of NGF on its target cells has arisen from the observation that inhibitors of protein methylation (which act indirectly via the blockade of S-adenosyl homocysteine hydrolase, resulting first in a build-up of S-adenosyl homocysteine, and then in a product inhibition of protein methylation) very specifically block both the short- and long-term effects of NGF in PC 12 cells (Seeley et al. 1984), bovine adrenal medullary cells (Acheson and Thoenen 1987) and chick sympathetic neurons (Acheson et al. 1986). The specificity of the effect of the inhibitors used was demonstrated by the fact that in PC 12 cells the responses to epidermal growth factor were not affected (Seeley et al. 1984).

In cultured bovine adrenal medullary cells both the NGF-mediated activation and induction of tyrosine hydroxylase was blocked, but not that produced by cAMP. In chick sympathetic neurons the survival effect of NGF was blocked, whereas the effect of 35 mmol/l potassium, which has a similar survival effect as NGF (Wakade et al. 1983), was not affected by the methyltransferase inhibitors (Acheson et al. 1986). The specificity of the effects of the inhibitors was further shown by the observation that the NGF-mediated changes in protein phosphorylation were blocked, whereas those mediated by 35 mmol/l potassium were not consistently affected. In particular, the methyltransferase inhibitors abolished the NGF-mediated, but not the 35 mmol/l potassium-mediated dephosphorylation of a 70 kD protein. In the absence of inhibitors both high potassium and NGF resulted in an identical dephosphorylation of this protein (Acheson et al. 1986). In spite of the high selectivity of the blocking action of inhibitors of methyltransferase on NGF-mediated effects, it seems unlikely that there is a causal relationship between the blockade of protein methylation and the selective blockade of NGF-mediated effects for the following reasons. In contrast to bacterial systems, protein methylation in eukaryotic cells has not yet been shown to serve a regulatory function. Indeed, the eukaryotic protein carboxylmethylating enzyme has been shown to affect only modified proteins, in which isoaspartate/D-aspartate residues are substrates (see Aswad 1984; O'Connor et al. 1984; Clarke 1985). Moreover, the substoichiometric nature of decarboxylmethylation of proteins such as calmodulin in intact or semi-purified systems (Johnson et al. 1985) also speaks against a regulatory function for this post-translational modification. The conversion of normal aspartate residues to isoaspartate which results in a stoichiometric carboxylmethylation (Johnson et al. 1985) suggests that carboxylmethylation serves as a tag for protein degradation or repair. Nevertheless the selectivity of the blockade of NGF's effect by inhibitors of protein methylation represents an attractive tool for future investigations on NGF's signalling mechanism.

3 NGF in the Central Nervous System

3.1 Identification of NGF-Responsive Neurons in the Central Nervous System

In the peripheral nervous system catecholaminergic cells, such as sympathetic neurons and adrenal chromaffin cells, together with neural crest-derived sensory neurons, represent the mean target cells of NGF (see Levi-Montalcini and Angeletti 1968; Greene and Shooter 1980; Thoenen and Barde 1980). Accordingly, the initial investigations in the central nervous system were focussed on catecholaminergic, i.e. dopaminergic and adrenergic neurons. However, the results of intraventricular and stereotactic intracerebral injections of NGF were all negative, for neither the dopaminergic neurons of the substantia nigra nor the adrenergic neurons of the locus coeruleus responded with an induction of tyrosine hydroxylase (Konkol et al. 1978; Schwab et al. 1979), a characteristic response of peripheral catecholaminergic target cells to NGF. Conversely no effect was observed after intraventricular and intracerebral injection of anti-NGF antibodies (Konkol et al. 1978; Schwab et al. 1979). The conclusiveness of these latter results, however, is questionable in view of the poor penetration of antibodies into brain tissue (see below). In view of these negative results the question arose as to whether they could be explained by the absence of NGF receptors. The injection of ^{125}I-labelled NGF into the field of innervation of the neurons of the substantia nigra and locus coeruleus did not result in a specific retrograde labelling of the corresponding cells bodies as previously demonstrated in the periphery for all NGF-responsive neurons (see Thoenen and Barde 1980; Schwab and Thoenen 1983). That the local injection of ^{125}I-labelled NGF was technically competent and that the nerve terminals projecting from the locus coeruleus and the substantia nigra were reached by the stereotactic injection of NGF, was demonstrated by the fact that ^{125}I-labelled tetanus toxin and wheat germ agglutinin (macromolecules transported retrogradely by all peripheral and central neurons investigated so far), injected in an identical manner as ^{125}I-labelled NGF, were transported retrogradely to the corresponding cell bodies. The injection of ^{125}I-labelled NGF into one of the projection fields of the locus coeruleus, the hippocampus, resulted in an unexpected, but important observation. Instead of the expected retrograde transport to the cell bodies of the locus coeruleus, a labelling of neurons of the nuclei of the medial septum and the diagonal band of Broca was observed (Schwab et al. 1979). These neurons projecting to the hippocampus had been suspected to be cholinergic in view of their positive reaction to acetylcholinesterase. Later, their cholinergic identity was proven with anti-choline acetyltransferase antibodies (Sofroniew et al. 1982; Eckenstein and Sofroniew 1983; Levey et al. 1983). In subsequent, more extensive investigations it was demonstrated that not only the hippocampal, but also the cortical projections of the cholinergic neurons in the basal forebrain nuclei showed specific uptake and retrograde axonal transport of ^{125}I-labelled NGF (Seiler and Schwab 1984). However, specific retrograde tracing with ^{125}I-labelled NGF is not suitable for identifying interneurons expressing NGF receptors. An important adjunct to the retrograde labelling procedure is the recently developed autoradiographic ^{125}I-NGF receptor binding procedure for cryostate sections of adult rat brains (Richardson et al. 1986; Raivich and Kreuzberg 1987). These studies demonstrated, in a confirmation of the retrograde labelling procedure, specific NGF binding by neurons of the medial septal nucleus, the diagonal band of Broca and the basal nucleus of

Meynert. Moreover, there was also labelling of a relatively sparse subpopulation of randomly distributed neurons in the striatum. In this context it is essential to note that cholinergic interneurons in the hippocampus and in the mesencephalic cortex were not labelled by the same procedure. This also holds for the neurons of the cholinergic motor nuclei of the brain stem, which were also not labelled. However, in the brain stem, specific NGF binding sites were located in a number of groups of neurons of the reticular formation, the dorso-lateral lemniscus and the cochlear nuclei. These labelled neurons in the brain stem are predominantly non-cholinergic (Raivich and Kreutzberg 1987). Their projection fields can only be deduced from available neuroanatomical information, since the autoradiographic receptor binding procedure only labels cell bodies, and not neuronal processes. Thus, this procedure is suitable for identifying interneurons (in contrast to the retrograde labelling procedure), although it does not provide information on the field of projection of the labelled neuronal cell bodies.

In summary, the predominant population of neurons expressing NGF receptors in the forebrain are cholinergic, whereas in the brain stem it is predomiantly the non-cholinergic neurons which seem to express NGF receptors. The functional role of the latter remains to be established.

3.2 Induction of Choline Acetyltransferase (ChAT) by Intraventricular Administration of NGF

Given impetus by the observation that the cholinergic neurons of the basal forebrain nuclei express NGF receptors (reflected by a specific uptake and retrograde axonal transport of ^{125}I-labelled NGF), the effect of intraventricularly injected NGF on ChAT levels in the basal forebrain nuclei and their fields of innervation has been studied. In newborn animals an increase in ChAT activity in hippocampus, medial septum and cortex (Gnahn et al. 1983) and in the striatum (Mobley et al. 1985) was observed. In adult animals a statistically significant increase in the ChAT activity was found only after repetitive injections of NGF over 4 weeks (Gnahn et al. 1983; Hefti et al. 1984). NGF was also found to induce ChAT in aggregate cultures of total fetal brain (Honegger and Lenoir 1982), and in explants and dissociated cultures of septum and striatum (Hefti et al. 1985a; Martinez et al. 1985). In several of these studies acetylcholinesterase was used as a histochemical cholinergic marker. Despite the caveat that this enzyme is generally an unreliable cholinergic marker, being expressed by both non-cholinergic neurons and even non-neuronal tissues, in the basal forebrain nuclei there is a very good correlation between the histochemical reaction of these neurons for acetylcholinesterase and their immunohistochemical identification as cholinergic neurons with specific ChAT antibodies (see Eckenstein and Sofroniew 1983; Levey et al.

1983; Cuello and Sofroniew 1984). In the nucleus basalis of Meynert the correlation is complete, and in the medial septum and the diagonal band of Broca only about 10% of the neurons which are acetylcholinesterase-positive are not stained by ChAT antibodies (see Eckenstein and Sofroniew 1983).

3.3 Effects of Intraventricular and Intracerebral Injection of Anti-NGF Antibodies

Although the demonstration of the presence of NGF receptors in the plasma membrane of a specific population of neurons, the uptake and retrograde axonal transport of NGF by these neurons and their specific biochemical response to exogenous NGF is certainly suggestive, it is not definite proof for a physiological function of NGF in the brain. In the peripheral nervous system the effects of anti-NGF antibodies played an essential role in establishing the physiological function of NGF (see Levi-Montalcini and Angeletti 1968; Thoenen and Barde 1980). As mentioned above, the elucidation of the effects of NGF antibody administration and the interruption of the retrograde axonal transport were the essential arguments for a physiological role of NGF in the peripheral sympathetic and sensory nervous system as a retrograde neurotrophic messenger (see Thoenen and Barde 1980; Schwab and Thoenen 1983; Otten 1984; Johnson et al. 1986). In the central nervous system the situation is not as straightforward and unambiguous as in the peripheral nervous system, both with respect to results obtained and their interpretation. In analogy to the peripheral nervous system it would be expected that the administation of anti-NGF antibodies would result in a decrease of the number of cholinergic neurons in the basal forebrain nuclei during a restricted period of their ontogenetic development and also in a reduction of their ChAT levels when they are fully differentiated. These changes would correspond to the reduction of the levels of enzymes involved in catecholamine synthesis and in a reduction of neuron-specific peptides in the peripheral sympathetic and sensory nervous system (see Thoenen and Barde 1980; Schwab and Thoenen 1983; Otten 1984). To carry the analogy further, because the function of NGF as a survival factor comes into play just as the NGF-responsive nerve fibers reach their target tissues, it would be expected that the cholinergic neurons of the basal forebrain nuclei would be particularly sensitive to NGF deprivation in the early postnatal period, when their connections in neocortex and hippocampus are forming (Angevine 1965; Matthews et al. 1974; Nadler et al. 1974; Sorimachi and Kataoka 1975; Zimmer and Haug 1978; Crutcher 1982; Milner et al. 1983; Nicoll 1985). However, both intraventricular and intracortical injection of polyclonal anti-NGF antibodies from birth to the 7th postnatal day had no effect on ChAT levels in the hippocampus, cortex and septum (Gnahn et al. 1983). Additonal efforts to influence ChAT levels in basal forebrain neurons by intraventricular injection of affinity-purified Fab-fragments of anti-NGF antibodies every second day from birth to the 14th postnatal day were also without effect (Thoenen et al. 1987a). The Fab-fragments used were shown to block the biological activity of NGF in vitro and, according to their size, should better penetrate into the brain tissue than intact IgG antibodies.

In contrast to these negative effects of the injection of anti-NGF antibodies in the early postnatal period, the injection of anti-NGF antibodies into rat fetuses at embryonic day 15.5 (E15.5) did lead to a significant decrease of ChAT levels in septum, hippocampus, and the nucleus basalis region (but not in the striatum) when levels were determined 6 weeks after birth (Otten et al. 1985). This effect is puzzling, since at the time of the antibody injection the majority of the cholinergic neurons of the basal forebrain nuclei have not yet been born (Bayer 1979a, b; 1985). Thus, two alternative interpretations of these effects have to be considered: a) at the time of the injection (E15.5) the penetration of the anti-NGF antibodies into the brain tissue, even after systemic injection, is better than after local injection into the brain tissue in the early postnatal period. The antibodies may possibly be preserved in the critical brain regions until the NGF-dependent period of NGF-responsive neurons starts in the early postnatal period; b) it is also possible that the cholinergic neurons of the basal forebrain nuclei depend on NGF for survival at an earlier period of their development than expected from analogy to the NGF-responsive neurons in the peripheral nervous system. Accordingly, in later periods of the development NGF might not even be essential for the maintenance of their specialized function. Possibly other kinds of trophic molecules could be responsible for the maintenance of ChAT levels (Appel et al. 1987).

However, firm conclusions from the negative results of the injection of antibodies in the postnatal period are not wholly justified, since they could also be explained by an insufficient penetration of the antibodies. Indeed, electron microscopic studies using a horseradish peroxidase-IgG conjugate at a ratio of 1:1 showed a very poor penetration into the surrounding brain tissue after intracerebral injection in newborn rats (Thoenen et al. 1987a). In these experiments it should be taken into account that the coupling product is about 200 kD and, thus, the penetration is even poorer than that of IgGs alone. Evidence for the poor penetration of (uncoupled) anti-NGF antibodies into the brain tissue is also provided by a recent report of Springer and Loy (1985), who demonstrated that the ingrowth of (peripheral) sympathetic fibers into the hippocampus after fimbria lesion was abolished after local injection of anti-NGF antibodies. However, the inhibitory effect was confined to the immediate vicinity of the injection site. At a distance of even only 1 mm from the injection site, the reactive ingrowth of sympathetic fibers was not impaired.

3.4 Regional Differences in the Distribution of NGF and its mRNA

In view of the difficulties in interpreting the results obtained after intraventricular and intracerebral injection of anti-NGF antibodies, more information was necessary to give an unequivocal answer to the question of whether NGF does play a physiological role in the central nervous system. In the periphery a further strong argument supporting the physiological function of NGF was the positive correlation between the levels of NGF and its mRNA in target tissues of NGF-responsive neurons and the density of their innervation (see above). Indeed, in the central nervous system a remarkably good correlation between the NGF and mRNA^{NGF} levels and the density of innervation by cholinergic neurons of the basal forebrain has also been demonstrated (Korsching et al. 1985; Korsching 1986; Shelton and Reichardt 1986a; Whittemore et al. 1986). NGF levels comparable to those observed in relatively densely innervated peripheral tissues were found in regions innervated by neurons of the basal forebrain nuclei and the regions containing their cell bodies, i.e. the hippocampus, neocortex, olfactory bulb, on the one hand, and medial septal nucleus, diagonal band of Broca, nucleus basalis of Meynert, on the other hand (Korsching et al. 1985; Korsching 1986; Whittemore et al. 1986). In this context it is of particular interest that the NGF levels in the dentate gyrus and the CA3-CA4 region of the hippocampus were 2- to 3-fold higher than in the CA1-CA2 region, demonstrating that also within discrete regions of the hippocampus the NGF levels reflected the density of cholinergic innervation (Korsching et al. 1985). NGF levels in the region of the basal forebrain nuclei were not as high as in the peripheral sympathetic ganglia (Korsching and Thoenen 1983a; Heumann et al. 1984). However, it has to be taken into account that the basal forebrain nuclei do not consist of an uniform population of neurons, and that the cholinergic neurons represent only a relatively small fraction of the total cell population (Eckenstein and Sofroniew 1983; Rye et al. 1984).

Brain regions not involved in the cholinergic basal forebrain system contain considerably lower NGF levels (Korsching et al. 1985; Whittemore et al. 1986). However, it could well be that, hidden within the relatively gross subdivisions so far assayed, there are more marked differences, as for instance in the subregions of the hippocampus (Korsching et al. 1985; Korsching 1986). It is interesting to note that the cerebellum, which has no known cholinergic input, has relatively high NGF levels (Korsching et al. 1985) and it remains to be established whether they reflect an innervation by non-cholinergic NGF-responsive neurons possibly located in the brain stem (Raivich and Kreutzberg 1987). In the striatum, which also contains NGF-responsive cholinergic neurons (Mobley et al. 1985), the NGF levels are particularly low (Korsching et al. 1985) and the levels of mRNA^{NGF} determined by Shelton and Reichardt (1986a) were only about 1/10 of those determined in the hippocampus. It should also be noted that the injection of rat fetuses with anti-NGF antibodies did not result in a ChAT reduction in the striatum, in contrast to the cholinergic basal forebrain system (Otten et al. 1985).

If one compares the correlation between the NGF levels and the density of innervation by NGF-responsive neurons in the periphery with that in the central nervous system, it readily becomes apparent that the situation in the cholinergic basal forebrain system is directly comparable. However, in other brain regions, in particular the brain stem and cerebellum, the situation is less clear. The ^{125}I-NGF labelling of NGF receptors by autoradiographic procedures in tissue sections (Richardson et al. 1986; Raivich and Kreutzberg 1987) cannot yet be combined with an immunological analysis which would allow the identification of ^{125}I-NGF labelled neurons producing specific neuropeptides or enzymes involved in transmitter synthesis. This information then would allow us to analyze in a focussed manner the responsiveness of these neurons and allow the assessment of whether their NGF receptors are functional or not. That this consideration is not merely a theoretical one evolves from recent investigations of Davies et al. (1987 b). They demonstrated that NGF has no survival effect on trigeminal mesencephalic neurons in culture, although all of these cells express NGF receptors. That this lack of an effect by NGF is not due to inappropriate general culture procedures is shown by the fact that virtually all the trigeminal mesencephalic neurons survive with brain-derived neurotrophic factor (Davies et al. 1986). It is hoped that in-situ hybridization experiments with NGF receptor cDNA or cRNA probes can be combined with immunostaining of the same or adjacent tissue sections allowing the identification of those neurons which express NGF receptors. A first indication of NGF-responsive central peptidergic neurons arose from the work of Levi-Montalcini and Aloe (1985), who demonstrated an increase in somatostatin and substance P in the CNS of Xenopus laevis tadpoles after systemic administration of NGF.

3.5 Comparison Between Developmental Changes of NGF and ChAT in the Septo-Hippocampal System

In contrast to the peripheral nervous system, in the CNS the injection of anti-NGF antibodies has not yet indicated a definite physiological role for NGF. The neurons of the septo-hippocampal system have been shown to express NGF receptors (Schwab et al. 1979; Seiler and Schwab 1984; Richardson et al. 1986; Raivich and Kreutzberg 1987), to respond to NGF by ChAT induction (Gnahn et al. 1983) and to display within the hippocampus a close correlation between the density of cholinergic innervation and the levels of NGF (Korsching et al. 1985; Whittemore et al. 1986). This system therefore seemed to be particularly suitable for comparing the developmental time-course of changes in NGF and ChAT levels and to deduce possible causal relationships. Indeed, during the postnatal development of the hippocampus the time-courses of NGF and ChAT increases are well correlated including a rapid 3-fold increase between postnatal (P) days P12 and P14 in the rat (Auburger et al. 1987; Thoenen et al. 1987a). The increase in hippocampal NGF was preceded by a corresponding increase in $mRNA^{NGF}$. The developmental changes in hippocampal NGF levels were also closely reflected in the corresponding changes in the septum. This observation, together with the finding in adult animals that the relatively high NGF levels in the septum were accompanied by $mRNA^{NGF}$ levels below (Korsching et al. 1985; Whittemore et al. 1986), or at the detection limit of (Shelton and Reichardt 1986a), the assay further support the notion that the NGF levels in the septum result from retrograde axonal transport rather than from local synthesis. One aspect of the developmental changes of NGF and $mRNA^{NGF}$ in this postnatal period deserves special attention, namely that in the hippocampus the $mRNA^{NGF}$ increase preceded that of NGF by several days (Auburger et al. 1987; Thoenen et al. 1987a). This considerable time-lag between the increase in $mRNA^{NGF}$ and the subsequent increase in NGF is in contrast to the periphery. There, wherever rapid changes in $mRNA^{NGF}$ were observed, they were followed within hours by a corresponding increase in NGF levels. This was for instance the case for both the rat iris in culture (Barth et al. 1984; Heumann and Thoenen 1986) or during the developmental changes of NGF and $mRNA^{NGF}$ in the whisker pad of the mouse, where the increase in $mRNA^{NGF}$ was also immediately followed by a corresponding increase in NGF protein (Davies et al. 1987a). That $mRNA^{NGF}$ levels precede the corresponding increases in NGF levels for several days is unique to the septo-hippocampal system, and it has to be assumed that either the $mRNA^{NGF}$ is initially present in a non-translatable form, or that, in the hippocampus during the period of the most rapid $mRNA^{NGF}$ increase, the $mRNA^{NGF}$ level is not rate-limiting for the production of NGF. If for instance the proteolytic processing enzymes for the NGF precursor were rate-limiting, a build-up of NGF precursor molecules would take place. These molecules are not recognized by antibodies directed against mature NGF and they also seem to be biologically inactive (Wion et al. 1984; Dicou et al. 1986). In addition to the unusually long time-lag between the increase in $mRNA^{NGF}$ and the subsequent increase in NGF levels, the time-course of the perinatal development of hippocampal NGF and $mRNA^{NGF}$ showed additional unexpected features. The NGF levels in the hippocampus from E17 to P0 were remarkably high, they corresponded to 1/3 to 1/4 of adult hippocampal NGF levels (Auburger et al. 1987; Thoenen et al. 1987a). In contrast, the $mRNA^{NGF}$ levels were below the detection limit suggesting that in this phase of the development there is a build-up of NGF, based on a small quantity of $mRNA^{NGF}$, and a lack of retrograde axonal transport to remove it. This interpretation is based on the fact that at this developmental stage the septal cholinergic neurons have not yet reached the hippocampus (Crutcher 1982) and there can therefore be no removal of NGF by retrograde axonal transport. This explanation for the relatively high prenatal hippocampal NGF levels is also supported by the observation that there is a distinct decrease in NGF from P0 to P2 when the first cholinergic fibers reach the hippocampus (Crutcher 1982). The decrease in hippocampal NGF levels by retrograde axonal removal is also mirrored in the corresponding increase of NGF in the septum (Auburger et al. 1987; Thoenen et al. 1987a).

Because a reliable dissection of the hippocampus before E17 was not possible, NGF levels were determined in samples of the whole telencephalic region (which include both hippocampus and neocortex) between E14 and E18. Astonishingly, in all the samples the NGF levels came close to those of adult cortex (Korsching et al. 1985; Auburger et al. 1987). Thus, one is confronted with the situation that the NGF levels supporting the prospective cholinergic neurons of the basal forebrain nuclei have already stabilised before a major part of the innervating neurons are even born and before all of them have established their connection to the target region (Bayer 1979a, b, 1985; Crutcher 1982). This temporal separation between NGF production in target areas and the ingrowth of the corresponding neurons is in distinct contrast to the peripheral nervous system. For instance in the whisker pad of the mouse, where the analysis has been performed most carefully, the increase in mRNA^{NGF} and NGF occurs virtually concomitantly with the arrival of the corresponding neurites from the trigeminal ganglion into the target fields (Davies et al. 1987a). In view of this temporal dissociation between the formation of NGF in the prospective target areas of the cholinergic neurons of the basal forebrain nuclei and the arrival of their outgrowing axons one has to consider the possibility that during this developmental period NGF fulfills functions which are not related to the cholinergic neurons. It could well be that other neurons transiently express NGF receptors and depend on NGF, or that NGF has still other unknown functions in the rat telencephalon during this developmental stage.

The extension of the autoradiographic ^{125}I-NGF binding studies to tissue sections of embryonic rat brain, and the availability of nucleotide probes for the rat NGF receptor (Radeke et al. 1987) (which will allow in-situ hybridization studies) should help to settle this question in the near future. It has also to be taken into consideration that at this early developmental stage NGF could have a physiological role for non-neuronal cells.

3.6 Cellular Localization and Site of Synthesis of NGF in the Central Nervous System

In contrast to the relatively detailed knowledge about the cellular localization and site of synthesis of NGF in the periphery, corresponding information for the central nervous system is at best fragmentary and controversial. The cellular localization of NGF in embryonic (Ayer-Lelievre et al. 1983; Ebendal et al. 1985) and adult brain (Whittemore et al. 1986) is based essentially on immunohistochemical observations, from which it is impossible to decide whether immunoreactivity is localized in neurons or glial cells. One exception may be the fibrillary structures in the cerebral cortex of adult rats, which might correspond to the projections of NGF-responsive cholinergic neurons (Whittemore et al. 1986). Unfortunately, no photographs are available of the cell bodies of the basal forebrain nuclei, where the highest immunoreactivity would be expected. A comparison with the cell bodies of peripheral sympathetic neurons would also be helpful because they contain by far the highest detectable NGF levels (Korsching and Thoenen 1983a, 1985b). If one compares the results of immunohistochemical experiments in the central nervous system, and in particular the fetal brain, with those in the periphery, the intensity of the signals in the central nervous system is generally much higher. For example, in the iris, and particularly the denervated iris, in distinct contrast to the high NGF levels (Ebendal et al. 1980, 1983; Korsching and Thoenen 1983a; Barth et al. 1984) the immunohistochemical reaction was very weak (Ayer-Lelievre et al. 1983; Rush 1984; Ebendal et al. 1985; Finn et al. 1986). This discrepancy might be due to the fact that the fixation conditions for the central nervous system were optimal for the visualization of NGF whereas those for the iris were poor. A similar explanation could be offered for the more intense staining seen in the fetal (Ayer-Lelievre et al. 1983; Ebendal et al. 1985) as compared to the adult brain (Whittemore et al. 1986). In the fetal rat brain immunoreactivity was found in many regions. Particularly strong staining was found in the ventral medulla oblongata and pons, posterior parts of the tectum, parts of the frontal cortex and the olfactory bulb. Areas with barely detectable immunofluorescence included the mesencephalic nigral region and the hypothalamus. Unfortunately, no quantitative data are available which would allow a comparison of the quantities of NGF determined by reliable enzyme immunoassays with these immunohistochemical observations. However, if one compares the immunohistochemical data from fetal rat brain (Ayer-Lelievre et al. 1983; Ebendal et al. 1985) with the data available for the regional distribution of NGF in the adult brain (Korsching et al. 1985; Whittemore et al. 1986) no obvious correlation is found. Of particular interest is the observation that in both fetal (Auburger 1987) and adult (Korsching and Thoenen 1985a) spinal cord the NGF levels were below the detection limit of a sensitive two-site enzyme immunoassay. In contrast, in the spinal cord of fetal rats there was a high general immunoreactivity, particularly intense in the ventral grey matter (Ayer-Lelievre et al. 1983; Ebendal et al. 1985).

The discrepancy between the relatively low NGF levels in the central nervous system as compared to the densely innervated organs in the periphery, and the inverse situation with respect to the immunohistochemical observations are disturbing. In particular, the puzzling discrepancy between the immunohistochemical findings in the spinal cord and the assayed NGF levels raises doubts about the specificity of the immunohistochemical observations. Although affinity-purified antibodies were used (Ayer-Lelievre et al. 1983; Ebendal et al. 1985; Whittemore et al. 1986), this method would not exclude the possibility that an antibody against a contaminant was responsible for the immunohistochemical staining. It has therefore become essential to compare that histochemical results obtained with affinity-purified antibodies with those obtained with monoclonal antibodies directed against different epitopes of mature NGF, and, if possible, against its precursor. It should also be remembered that antibodies recognizing mature NGF do not recognize the NGF-precursor and vice-versa (Dicou et al. 1986). This NGF-inherent complication, when added to the general problems of immunohistochemistry creates serious uncertainty as to which antigen is actually being recognized in fixed tissues. A further sobering example of such problems is provided by epidermal growth factor. In this case a highly specific regional distribution of epidermal growth factor-reactive material was demonstrated in the rodent brain by immunohistochemistry (Fallon et al. 1984), and yet epidermal growth factor itself could not be detected with a sensitive and specific two-site enzyme immunoassay (Probstmeier and Schachner 1986).

Very recently Rennert and Heinrich (1986) reported on the localization of mRNA^{NGF} in the stratum granulosum of the dentate gyrus and the stratum pyramidale of the hippocampus by in situ-hybridization. The disposition of the label (P^{32}) was taken to indicate that mRNA^{NGF} is localized in neurons. However, their experimental procedure is open to criticism. The authors used a P^{32}-labeled cRNA probe whose relatively large size (460 bases) is not optimal for in situ-hybridization. As proof for the specificity of the hybridization signal, the authors showed that *a)* pre-digestion with ribonuclease abolished the autoradiographic signal and *b)* pre-incubation with excess unlabelled cRNA reduced the autoradiographic signal over the gyrus dentatus and the stratum pyramidale. However, these controls are insufficient to demonstrate the specificity of their hybridization signal. They merely demonstrate that the P^{32}-labeled cRNA probe reacted with structures containing large quantities of RNA. RNA probes of strand and counterstrand have to be compared, and the stringency of the hybridization-procedure increased so that the signal arising from the strand corresponding to the mRNA^{NGF} disappears before the signal of the complementary strand. Using this approach our laboratory obtained exactly the same localization as Rennert and Heinrich (1986) with S^{35}- and P^{32}-labeled single-stranded RNA probes of NGF. However, the signal proved to be non-specific, because it disappeared for both strands at the same stringency (unpublished observation). In this context, the observations of Schalling et al. (1986) with cRNA probes for tyrosine hydroxylase are also very instructive. They demonstrated that the same structures in the hippocampus were strongly labeled by strand and counterstrand of RNA probes of tyrosine hydroxylase. Thus, it seems likely that the labeling of cells in the hippocampus may be an artifact reflecting the high RNA content of these cells. This interpretation is further supported by the fact that in the hippocampus $mRNA^{NGF}$ levels are distinctly lower than, for example, in the rat iris which contains about $1-2$ copies of $mRNA^{NGF}$ per cell (Bandtlow et al. 1987). The signal arising after long exposure times just reached the limit of detectability (Bandtlow et al. 1987). Using the same experimental procedure for the hippocampus as for the iris we were not able to obtain a specific signal over the hippocampus. This is not surprising in view of the very low $mRNA^{NGF}$ copy numbers present there.

In agreement with the earlier observations of Lindsay (1979) that NGF is produced by rat astrocytes in culture, in primary cultures of various brain regions we could detect specific hybridization signals only over astrocytes (unpublished observations). It has not yet been determined whether the labeled cells represent a specific subpopulation. However, these results also have to be considered with great caution, because, as was impressively demonstrated in the case of laminin (Liesi et al. 1983), the expression of genetic information in cultured astrocytes can differ significantly from that in situ.

3.7 The Role Played by Fimbria Lesion Experiments in the Elucidation of a Potential Physiological Function of NGF in the Central Nervous System

About the same time that it was demonstrated that central catecholaminergic neurons do not respond to NGF (Konkol et al. 1978; Schwab et al. 1979; Olson et al. 1979; Dreyfus et al. 1980) and do not express NGF receptors (Schwab et al. 1979), but that NGF is transported specifically from the hippocampus to septal cholinergic neurons (Schwab et al. 1979), several laboratories reported that lesioning the septo-hippocampal pathway resulted in an ingrowth of perivascular sympathetic fibers into the hippocampus (Loy and Moore 1977; Stenevi and Björklund 1978; Crutcher et al. 1979; Crutcher and Davis 1981). It was also demonstrated that only a fimbrial transection could trigger this reactive ingrowth of peripheral sympathetic nerve fibers. In contrast, interruption of other afferent pathways to the hippocampus, in particular the entorhinal pathways and the adrenergic input from the locus coeruleus had no effect (Björklund and Stenevi 1981). Later, it was demonstrated that this reactive ingrowth occurred predominantly into the dentate-

CA3 region, and that the transection enhanced the survival of superior cervical ganglia transplanted into the denervated hippocampus of newborn rats (Gage et al. 1984b). In more recent experiments Collins and Crutcher (1985) investigated the effect of fimbria lesion on the fiber outgrowth-promoting activity of tissue culture media conditioned over hippocampus sections. These workers used chick sympathetic ganglion explants as an assay system, and exposed the ganglia to the conditioned media for 48 hours. The fiber outgrowth activity in the conditioned media of CA1 sections of non-lesioned hippocampi was about half that in the CA3-dentate region. Lesion of the fimbria a week before the assay resulted in a doubling of the fiber outgrowth activity in CA1-conditioned media, and in an about 50% increase in the dentate CA3 region. Under these experimental conditions a considerable part of the fiber outgrowth activity could be neutralized by anti-NGF antibodies. Although the differences obtained in this biological assay system between the fiber outgrowth-promoting activity of the intact dentate-CA3 region and the CA1-CA2 region is smaller than the difference determined in these two regions by the two-site enzyme immunoassay (Korsching et al. 1985), the maximal increase after fimbria lesion is about the same as that determined by the enzyme immunoassay for both the entire (Korsching et al. 1986) and the ventral and dorsal parts of the hippocampus (Gasser et al. 1986). Although these experiments support the concept that the ingrowth of sympathetic fibers into the hippocampus after fimbria lesion is at least partially mediated by NGF, the most crucial and convincing evidence was provided by the recent experiments of Springer and Loy (1985). These workers demonstrated that local injection of anti-NGF antibodies into the hippocampus abolished the reactive ingrowth of sympathetic fibers after fimbrial lesion, although the effect of the antibody injection was limited to the immediate vicinity of the injection site due to the limited diffusion of the antibodies (see above). That the hippocampus also produces additional neurotrophic factors is supported by the observation that extracts of hippocampus contain activity which can promote fiber outgrowth from chick ciliary neurons which cannot be abolished by anti-NGF antibodies (Crutcher and Collins 1982; see also Ojika and Appel 1984). Additional evidence for the production of neurotrophic molecules distinct from NGF come from the observations of Björklund and Stenevi (1981) that after fimbria lesion there is not only a reactive ingrowth of perivascular (peripheral) sympathetic nerve fibers, but also an enhanced ingrowth of adrenergic fibers from the locus coeruleus. These adrenergic neurons have previously been shown neither to express NGF receptors nor to respond to NGF (Konkol et al. 1978; Schwab et al. 1979; Dreyfus et al. 1980). In contrast to the very marked NGF increases after denervation in the periphery (Ebendal et al. 1980, 1983; Barth et al. 1984; Korsching and Thoenen 1985b), the NGF increase in the hippocampus resulting from fimbria lesion is only about 50% (Korsching et al. 1986; Gasser et al. 1986). This increase can easily be explained by an elimination of the retrograde axonal transport. Moreover, it is noteworthy that the NGF increase does not occur immediately, as for example, after the administration of 6-hydroxydopamine in the periphery (Korsching and Thoenen 1985b). After fimbria lesion maximal NGF levels were attained either between 3 days to 1 week (Gasser et al. 1986) or after 2 weeks (Korsching et al. 1986). This delay in the small increase in NGF levels after interruption of the cholinergic input may be explained by a slower degeneration of axons distal to a lesion in the central as compared to the peripheral nervous system. The time difference in obtaining the maximal NGF levels after fimbria lesion between the experiments of Gasser et al. (1986) (3 days to 1 week) and of Korsching et al. (1986) (2 weeks) may be due to the fimbria lesion procedure used. In this context a puzzling observation of Gasser et al. (1986) must be mentioned. These workers found that, after fimbria lesion, instead of the expected decrease in septal NGF levels (see above), a marked increase occurred. Because fimbria lesion leads to an extensive degeneration of neurons in the septum and the Broca band (which would be expected to result in a decrease in NGF accumulated in cholinergic neurons by retrograde transport; Hefti 1986; Williams et al. 1986), it has to be assumed that this marked increase (400%), reaching a maximum after 1 week, results from a reactive local production. This enhanced production might result from a direct mechanical lesion (the site of transection of the fimbria is very close to the septum) leading to a reactive gliosis, accompanied by an enhanced NGF production by these glial cells. A similar reactive gliosis could also result as a response to the degeneration of septal cholinergic neurons after fimbria lesion (see Hefti 1986; Williams et al. 1986). The assumption of an enhanced local NGF synthesis is also supported by the observation that, after fimbria lesion, in spite of the resulting degeneration of about 50% of the neurons in the septum, there is no corresponding decrease in the ChAT activity, as would be expected as a result of the degeneration of the cholinergic neurons. After a small reduction in ChAT levels 3 days postlesion, the ChAT levels return to normal or even to a slightly elevated level (Gasser et al. 1986). In view of the demonstrated degeneration of cholinergic neurons (Hefti 1986; Williams et al. 1986) one has to assume that the local reactive production of NGF results in an augmented production of ChAT in the surviving neurons. The solution to this puzzling observation of the marked increase in NGF levels in the septum after fimbria lesion and the lack of a reduction of ChAT and acetylcholinesterase levels, in spite of an extensive degeneration of the cholinergic septal neurons, may be found by investigating whether local mechanical lesions in the brain lead to an enhanced NGF synthesis in the surrounding areas.

The marked ingrowth of sympathetic nerve fibers into the hippocampus after the elimination of the septal cholinergic input is very surprising in view of the relatively small increases in NGF levels determined by two-site enzyme immunoassays (Korsching et al. 1986; Gasser et al. 1986) or estimated by a biological assay for fiber outgrowth (Collins and Crutcher 1985). It is hard to conceive how a 50% increase in NGF levels could result in such a strong fiber ingrowth. Either one has to assume that additional neurotrophic factors are produced which act on sympathetic nerve fibers (which, however, should be detected in bioassays), or that the NGF effect is potentiated. Indeed, it has been demonstrated in vitro that the extracellular matrix glycoprotein laminin potentiates very markedly fiber outgrowth and survival effect of NGF by about one order of magnitude (Edgar et al. 1984). It has to be emphasized that laminin has no neurotrophic effect on its own; it can only enhance the effect of NGF and other neurotrophic molecules (Barde et al. 1987; Thoenen et al. 1987b). Thus, it could well be that the cholinergic denervation of the hippocampus leads to the formation of such potentiating molecules. For example, it has been demonstrated that after mechanical brain lesion or cytotoxic degeneration of neurons, astrocytes start to produce laminin (Liesi et al. 1984). Laminin production by astrocytes after various brain lesions is also suggested by the continuous production of laminin by astrocytes in tissue culture (Liesi et al. 1983).

3.8 Effects of NGF After Fimbria Lesion

In adult rats even prolonged repetitive injections (4 weeks) of NGF resulted only in a very small (15%) increase in septal hippocampal and cortical ChAT levels (Gnahn et al. 1983). After partial fimbria lesion Hefti et al. (1984) demonstrated that this borderline effect of NGF treatment in adult animals was markedly enhanced. The ChAT increase on the lesioned side amounted to 60%. In agreement with the NGF effect in newborn animals, where NGF produces a marked ChAT increase in the neocortical and septal areas, but no increase in acetylcholinesterase (Gnahn et al. 1983), there was no discernible effect of NGF in adult animals after partial fimbria lesion on acetylcholinesterase levels, either on the intact or the lesioned side. This result argues against a sprouting of intact cholinergic fibers, but favours the assumption of an induction of ChAT or (see below) the enhanced survival of cholinergic neurons which would otherwise die. In recent experiments Hefti (1986) has found that the complete transection of fimbria in adult rats resulted in a loss of neuronal cell bodies in the medial-septal nucleus and in the vertical limb of the diagonal band of Broca. Furthermore, in these same nuclei there was also a reduction in the number of cholinergic neurons. On the lesioned side of the medial-septal nucleus and the vertical limb of the Broca band the cholinergic cell bodies were reduced by 50% as compared to the intact contralateral side. Repetitive intraventricular injection of NGF through an implanted cannula virtually abolished the degeneration of neurons in these regions (Hefti 1986).

Independently, Williams et al. (1986) have made similar observations. These workers also came to the conclusion that the non-cholinergic neurons in these regions were also affected. In the studies of Williams et al. (1986) the effect of fimbria lesion and NGF treatment were separately investigated in the medial-septal nucleus and the vertical limb of the diagonal band of Broca. The continuous administration of NGF by an infusion pump rescued at least 50% of the cholinergic and non-cholinergic neurons which would otherwise degenerate after fimbria lesion in the medial-septal region and even up to 100% in the vertical limb of the diagonal band of Broca. It remains to be established whether the rescue of the non-cholinergic neurons by NGF is due to a direct action of NGF, or whether it results from an indirect effect mediated by the cholinergic neurons. The latter assumption seems to be correct, since it has been demonstrated that NGF receptors are present only on cholinergic neurons, at least in the human basal forebrain (Hefti et al. 1986). At least part of the non-cholinergic population could be GABAergic, since it has been demonstrated that glutamic acid decarboxylase-positive neurons project from the septum to the hippocampus (Köhler et al. 1984). In this context the recent observation of Mobley et al. (1986) is of interest that in newborn rats NGF does not influence the levels of glutamic acid decarboxylase in regions where NGF increased the ChAT levels.

The sparing effect of NGF after fimbria lesion does not eo ipso prove that the infused NGF replaces an interrupted natural supply of NGF. It might also be replacing other endogenous neurotrophic molecules. However, it does indicate that an augmented supply of NGF to damaged neurons (whatever the cause) has a beneficial pharmacological effect and opens up interesting possibilities for the treatment of Alzheimer's disease (see below).

4 NGF and Alzheimer's Disease: Possible Causal Relationships and Therapeutic Implications

Alzheimer's disease is characterized by a progressive loss of memory and other cognitive functions, which result in severe disability (see Hefti and Weiner 1986). The neuropathological correlates to these functional changes are the presence of paired helical filaments localized in neuritic plaques, and neurofibrillary tangles in neuronal cell bodies, the latter predominantly in the neocortex and hippocampus. These characteristic changes are associated with degeneration and/or atrophy of specific neuronal systems which synthesize biogenic amines and peptides such as acetylcholine, noradrenaline and somatostatin (see Perry et al. 1978; Willcock et al. 1983; Wisniewski and Merz 1983; Rossor et al. 1984; Price et al. 1985; Hefti and Weiner 1986). The loss and/or atrophy of cholinergic neurons (Coyle et al. 1983; Pearson et al. 1983;

Arendt et al. 1984) seems to be the most consistent neuropathological finding in Alzheimer's disease. The neurons affected are the cholinergic neurons of the basal forebrain nuclei which project to the hippocampus, neocortex and amygdala. Moreover, the impairment of acetylcholine synthesis is the earliest sign of the disease and is well correlated with cognitive impairments. This was demonstrated in bioptic specimens obtained from diagnostic craniotomies of young Alzheimer patients (Francis et al. 1985). The involvement of the cholinergic system in the clinical manifestations of Alzheimer's disease is also supported by various experimental observations. For instance, in rats the interruption of the ascending cholinergic projections from the basal forebrain nuclei results in a marked reduction of memory and learning ability (Hefti et al. 1985b; Hepler et al. 1985; Will and Hefti 1985). These learning and memory deficits can be improved by either injecting NGF (Stein and Will 1983; Will and Hefti 1985) or by the transplantation of fetal cholinergic neurons into the hippocampus (Dunett et al. 1982; Low et al. 1982; Gage and Björklund 1986). Although in patients the therapeutic benefit of cholinomimetics remains a matter of debate (see Davies 1985; Collerton 1986; Hefti and Weiner 1986), at least in rats cholinomimetic (*i.e.* muscarinic) agonists do seem to have a beneficial effect, not only on the performance of animals with septo-hippocampal lesions, but also in subpopulations of aged rats with learning and memory deficits (Bartus et al. 1982). These deficits are also improved by implantation of fetal septal neurons into the hippocampus (Gage et al. 1984a). These experiments demonstrate that the cholinergic system of the basal forebrain nuclei seems to be strongly involved in the learning deficits of the animal models used. The available pathophysiological information from Alzheimer patients, and the complementary information from animal experiments, opens up interesting possibilities for the elucidation of the pathophysiological causes of Alzheimer's disease and potential new therapeutic approaches. The availability of cDNA probes for human NGF (Ullrich et al. 1983), the possibility of producing human NGF by biotechnological methods, the consequent production of specific antibodies against human NGF, and the development of a specific enzyme immunoassay, are all prerequisites for an experimental approach to the question whether Alzheimer's disease is actually associated with a deficit in the production of human NGF. If such a deficit were found, it would also be necessary to postulate that as well as a reduced production of NGF, there is also reduced production of other, unknown neurotrophic factors acting on populations of neurons, which are also affected by Alzheimer's disease, but which are not responsive to NGF. Very recently, during the preparation of this manuscript, Goedert et al. (1986) reported that they could not find a difference in the $mRNA^{NGF}$ in the brain of Alzheimer patients as compared to age-matched controls.

With respect to the therapeutic consequences, the benefits of NGF administration on learning deficits after experimental lesions of cholinergic systems suggest that, whatever the cause of the damage of the cholinergic neurons is, an increased availability of NGF for these neurons, either by exogenous application or by stimulation of endogenous production, could be substantial. Although the production of human NGF by biotechnological methods is in principle possible and would eliminate potential immunological pitfalls of a therapy with non-human NGF, such therapy would raise ethical problems, which will not be discussed here. Beyond the ethical problems, the intracerebral administration (Harbaugh 1986) of recombinant NGF is barely practical in view of the large, and continuously increasing number of future patients affected by Alzheimer's disease. Particular attention, however, should be paid to investigations aiming at the elucidation of the regulation of the synthesis of endogenous NGF, and the possibilities of enhance this synthesis by pharmacological procedures. Such a therapy seems to be particularly promising for early stages of the disease, when affected neurons are not yet irreversibly damaged and can be supported by an augmented supply of exogenously applied or endogenously produced NGF.

Acknowledgements. We are grateful to Victor Nurcombe and Yves-Alain Barde for critical reading and Ursula Grenzemann for the preparation of the manuscript. Part of the research presented was supported by the Deutsche Forschungsgemeinschaft, Grant Th 270/3-2

References

Acheson A, Thoenen H (1987) Both short and long-term effects of nerve growth factor on tyrosine hydroxylase in calf adrenal chromaffin cells are blocked by methyltransferase inhibitors. J Neurochem 48:1416–1424

Acheson A, Vogl W, Huttner WB, Thoenen H (1986) Methyltransferase inhibitors block NGF-regulated survival and protein phosphorylation in sympathetic neurons. EMBO J 5:2799–2803

Angevine J (1965) Time of neuron origin in the hippocampal region. In: Exp Neurol [Suppl 2], Academic Press, New York, pp 1–70

Appel SH, McNaman M, Smith RG, Voca K, Bostwick JR (1987) Cholinergic trophic factors from skeletal muscle and hippocampus. In: Cholinergic Mechanisms, vol 6, Cellular and Molecular Basis of Cholinergic Function. Ellis Horwood, Chichester (in press)

Arendt T, Bigl V, Tennstedt A, Arendt A (1984) Correlation between cortical plaque count and neuronal loss in the nucleus basalis in Alzheimer's disease. Neurosci Lett 48:81–85

Aswad D (1984) Stoichiometric methylation of porcine adrenocorticotropin by protein carboxyl methyltransferase requires deamidation of asparagine 25. J Biol Chem 259:10714–10721

Auburger G (1987) Demonstration von Nerve Growth Factor im Zentralnervensystem: Evidenz für eine physiologische Rolle für die cholinerg-magnozellulären Neuronen des basalen Vorderhirns. Thesis Technical University Munich

Auburger G, Heumann R, Hellweg R, Korsching S, Thoenen H (1987) Developmental changes of nerve growth factor (NGF) and its mRNA in the rat hippocampus: Comparison with choline acetyltransferase. Dev Biol 120:322–328

Ayer-LeLievre CS, Ebendal T, Olson L, Seiger A (1983) Localization of nerve growth factor-like immunoreactivity in rat nervous tissue. Medical Biology 61:296–304

Bandtlow CE, Heumann R, Schwab ME, Thoenen H (1987) Cellular localization of nerve growth factor synthesis by in situ-hybridization. EMBO J 6:891−899

Barde Y-A, Davies AM, Johnson JE, Lindsay RM, Thoenen H (1987) Brain-derived neurotrophic factor. Prog Brain Res 71:185−189

Bar-Sagi D, Feramisco JR (1985) Microinjection of the *ras* oncogene protein into PC12 cells induces morphological differentiation. Cell 42:841−848

Barth EM, Korsching S, Thoenen H (1984) Regulation of nerve growth factor synthesis and release in organ cultures of rat iris. J Cell Biol 99:839−843

Bartus RT, Dean RL, Beer B, Lippa AS (1982) The cholinergic hypothesis of geriatric memory dysfunction. Science 217:408−417

Bayer SA (1979a) The development of the septal region in the rat. I. Neurogenesis examined with ^{3}H-thymidine autoradiography. J Comp Neurol 183:89−106

Bayer SA (1979b) The development of the septal region in the rat. II. Morphogenesis in normal and x-irradiated embryos. J Comp Neurol 183:107−120

Bayer SA (1985) Neurogenesis of the magnocellular basal telencephalic nuclei in the rat. Int J Dev Neurosci 2:229−243

Björklund A, Stenevi U (1981) In vivo evidence for a hippocampal adrenergic neuronotrophic factor specifically released on septal deafferentation. Brain Res 229:403−428

Chandler CE, Parsons LM, Hosang M, Shooter EM (1984) A monoclonal antibody modulates the interaction of nerve growth factor with PC12 cells. J Biol Chem 259:6882−6889

Clarke S (1985) Protein carboxyl methyltransferases: distinct classes of enzymes. Ann Rev Biochem 54:479−506

Collerton D (1986) Cholinergic function and intellectual decline in Alzheimer's disease. Neuroscience 19:1−28

Collins F, Crutcher KA (1985) Neurotrophic activity in the adult rat hippocampal formation: regional distribution and increase after septal lesion. J Neurosci 5:2809−2814

Coyle JT, Price DL, de Long MR (1983) Alzheimer's disease: a disease of cortical cholinergic innervation. Science 219:1184−1189

Crutcher KA (1982) Development of the rat septohippocampal projection: a retrograde fluorescent tracer study. Dev Brain Res 3:145−150

Crutcher KA, Collins F (1982) In vitro evidence for two distinct hippocampal growth factors: basis of neuronal plasticity? Science 217:67−68

Crutcher KA, Davis JN (1981) Sympathetic noradrenergic sprouting in response to central cholinergic denervation. Trends Neurosci 4:70−72

Crutcher KA, Davis JN (1982) Target regulation of sympathetic sprouting in the rat hippocampal formation. Exp Neurol 75:347−359

Crutcher KA, Brothers L, Davis JN (1979) Sprouting of sympathetic nerves in the absence of afferent input. Exp Neurol 66:778−783

Cuello AC, Sofroniew MV (1984) The anatomy of the CNS cholinergic neurons. Trends Neurosci 7:74−78

Davies AM, Thoenen H, Barde Y-A (1986) Different factors from the central nervous system and periphery regulate the survival of sensory neurones. Nature 319:497−499

Davies AM, Bandtlow C, Heumann R, Korsching S, Rohrer H, Thoenen H (1987a) The site and timing of nerve growth factor (NGF) synthesis in developing skin in relation to its innervation by sensory neurones and their expression of NGF receptors. Nature 326:353−363

Davies AM, Lumsden AGS, Rohrer H (1987b) Neural crest-derived proprioceptive neurons express NGF receptors but are not supported by NGF in culture. Neuroscience 20:37−46

Davies P (1985) Is it possible to design rational treatment for the symptoms of Alzheimer's disease? Drug Dev Res 5:69−75

Dicou E, Lee J, Brachet P (1986) Synthesis of nerve growth factor mRNA and precursor protein in the thyroid and parathyroid glands of the rat. Proc Natl Acad Sci USA 83:7084−7088

Dreyfus CF, Peterson ER, Crain SM (1980) Failure of nerve growth factor to affect fetal mouse brain catecholaminergic neurons in culture. Brain Res 194:540−547

Dunnett SB, Low WC, Iversen SD, Stenevi U, Björklund A (1982) Septal transplants restore maze learning in rats with fornix-fimbria lesions. Brain Res 251:335−348

Ebendal T, Olson L, Seiger A, Hedlund K-O (1980) Nerve growth factors in the rat iris. Nature 286:25−28

Ebendal T, Olson L, Seiger A (1983) The level of nerve growth factor (NGF) as a function of innervation. A correlative radio-immunoassay and bioassay study of the rat iris. Exp Cell Res 148:311−317

Ebendal T, Lärkfors L, Ayer-LeLievre C, Seiger A, Olson L (1985) New approaches to detect NGF-like activity in tissues. In: Dumont JE, Hamprecht B, Nunez J (eds) Hormones and Cell Regulation, vol 9, Elsevier, Amsterdam, pp 361−375

Eckenstein F, Sofroniew MW (1983) Identification of central cholinergic neurons containing both choline acetyltransferase and acetyl-cholinesterase and of central neurons containing only acetylcholinesterase. J Neurosci 3:2286−2291

Edgar D, Timpl R, Thoenen H (1984) The heparin-binding domain of laminin is responsible for its effects on neurite outgrowth and neuronal survival. EMBO J 3:1463−1468

Fallon JH, Seroogy KB, Loughlin SE, Morrison RS, Bradshaw RA, Knauer DJ, Cunningham DD (1984) Epidermal growth factor immunoreactive material in the central nervous system: Location and development. Science 224:1107−1109

Finn PJ, Ferguson IA, Renton FJ, Rush RA (1986) Nerve growth factor immunohistochemistry and biological activity in the rat iris. J Neurocytol 15:169−176

Francis PT, Palmer AM, Sioms NR, Bowen DM, Davison AN, Esiri MM, Neary D, Smowden JS, Wilcock GK (1985) Neurochemical studies of early-onset Alzheimer's disease. Possible influence on treatment. Lancet 4:7−11

Gage FH, Björklund A (1986) Cholinergic septal grafts into the hippocampal formation improve spatial learning and memory in aged rats by an atropine-sensitive mechanism. J Neurosci 6:2837−2847

Gage FH, Björklund A, Stenevi U, Dunnett SB, Kelly PAT (1984a) Intrahippocampal septal grafts ameliorate learning impairments in aged rats. Science 225:533−536

Gage FH, Björklund A, Stenevi U (1984b) Denervation releases a neuronal survival factor in adult rat hippocampus. Nature 308:637−639

Gasser UE, Weskamp G, Otten U, Dravid AR (1986) Time course of the elevation of nerve growth factor (NGF) content in the hippocampus and septum following lesions of the septohippocampal pathway in rats. Brain Res 376:351−356

Gnahn H, Hefti F, Heumann R, Schwab M, Thoenen H (1983) NGF-mediated increase of choline acetyltransferase (ChAT) in the neonatal forebrain: Evidence for a physiological role of NGF in the brain? Dev Brain Res 9:45−52

Goedert M, Fine A, Hunt SP, Ullrich A (1986) Nerve growth factor mRNA in peripheral and central rat tissues and in the human central nervous system: Lesion effects in the rat brain and levels in Alzheimer's disease. Mol Brain Res 1:85−92

Gorin MD, Johnson EM Jr (1979) Experimental autoimmune model of nerve growth factor deprivation: Effects on developing peripheral sympathetic and sensory neurons. Proc Natl Acad Sci USA 76:5382−5386

Gorin PD, Johnson EM (1980) Effects of long-term nerve growth factor deprivation on the nervous system of the adult rat: an experimental autoimmune approach. Brain Res 198:27−42

Greene LA, Shooter EM (1980) The nerve growth factor: biochemistry, synthesis, and mechanism of action. Ann Rev Neurosci 3:353−402

Harbaugh RE (1986) Intracranial drug administration in Alzheimer's disease. Psychopharmacol Bull 22:106−109

Hagag N, Halegoua S, Viola M (1986) Inhibition of growth factor-induced differentiation of PC12 cells by microinjection of antibody to ras p21. Nature 319:680−682

Harper GP, Glanville RW, Thoenen H (1982) The purification of nerve growth factor from bovine seminal plasma. J Biol Chem 257:8541−8548

Harper GP, Barde Y-A, Edgar D, Ganten D, Hefti F, Heumann R, Naujoks KW, Rohrer H, Turner JE, Thoenen H (1983) Biological and immunological properties of the nerve growth factor from bovine seminal plasma: comparison with the properties of mouse nerve growth factor. Neuroscience 8:375−387

Hefti F (1983) Alzheimer's disease caused by a lack of nerve growth factor? Ann Neurol 13:109−110

Hefti F (1986) Nerve growth factor (NGF) promotes survival of septal cholinergic neurons after fimbrial transections. J Neurosci 6:2155−2162

Hefti F, Weiner WJ (1986) Nerve growth factor and Alzheimer's disease. Ann Neurol 20:275−281

Hefti F, Dravid A, Hartikka J (1984) Chronic intraventricular injections of nerve growth factor elevate hippocampal choline acetyltransferase activity in adult rats with septo-hippocampal lesions. Brain Res 293:305−311

Hefti F, Hartikka J, Eckenstein F, Gnahn H, Heumann R, Schwab M (1985a) Nerve growth factor (NGF) increases choline acetyltransferase but not survival or fiber outgrowth of cultured fetal spetal cholinergic neurons. Neuroscience 14:55−68

Hefti F, Hartikka J, Will B (1985b) Effects of nerve growth factor on cholinergic neurons of the rat forebrain. In: Will BE, Schmitt P, Dalrymple-Alford JC (eds) Brain plasticity, learning and memory. Plenum Press, New York, pp 495−504

Hefti F, Hartikka J, Salvatierra A, Weiner WJ, Mash DC (1986) Localization of nerve growth factor receptors in cholinergic neurons of the human basal forebrain. Neurosci Lett 69:37−41

Hendry JA (1980) Proteins of the nervous system. 2nd ed. Bradshaw RA, Schneider DM (eds) Raven Press, New York, pp 183−211

Hepler DJ, Olton DS, Wenk GL, Coyle JT (1985) Lesions in nucleus basalis magnocellularis and medial septal area of rats produce qualitatively similar memory impairments. J Neurosci 5:866−873

Heumann R, Thoenen H (1986) Comparison between the time course of changes in nerve growth factor (NGF) protein levels and those of its messenger RNA in the cultured rat iris. J Biol Chem 261:9246−9249

Heumann R, Schwab M, Thoenen H (1981) A second messenger required for nerve growth factor biological activity? Nature 292:838−840

Heumann R, Korsching S, Scott J, Thoenen H (1984) Relationship between levels of nerve growth factor (NGF) and its messenger RNA in sympathetic ganglia and peripheral target tissues. EMBO J 3:3183−3189

Heumann R, Korsching S, Bandtlow C, Thoenen H (1987) Changes of nerve growth factor synthesis in non-neuronal cells in response to sciatic nerve transection. J Cell Biol 104:1623−1631

Hogue-Angeletti R, Bradshaw RA (1971) Nerve growth factor from the mouse submaxillary gland: amino acid sequence. Proc Natl Acad Sci USA 68:2417−2420

Honegger P, Lenoir D (1982) Nerve growth factor (NGF) stimulation of cholinergic telencephalic neurons in aggregating cell cultures. Dev Brain Res 3:229−239

Hopp TP, Woods KR (1981) Prediction of protein antigenic determinants from amino acid sequences. Proc Natl Acad Sci USA 78:3824−3828

Hosang M, Shooter EM (1985) Molecular characteristics of nerve growth factor receptors on PC12 cells. J Biol Chem 260:655−662

Johnson BA, Freitag NE, Aswad DW (1985) Protein carboxyl methyltransferase selectively modifies an atypical form of calmodulin. Evidence for methylation at deamidated asparagine residues. J Biol Chem 260:10913−10916

Johnson EM Jr, Gorin PD, Brandeis LD, Pearson J (1980) Dorsal root ganglion neurons are destroyed by exposure in utero to maternal antibody to Nerve Growth Factor. Science 210:916−918

Johnson EM Jr, Rich KM, Yip HK (1986) The role of NGF in sensory neurons in vivo. Trends Neurosci 1:33−37

Köhler C, Chan-Palay V, Wu J-Y (1984) Septal neurons containing glutamic acid decarboxylase immunoreactivity project to the hippocampal region in the rat brain. Anat Embryol (Berl) 169:41−44

Konkol RJ, Mailman RB, Bendeich EG, Garrison AM, Mueller RA, Breese GR (1978) Evaluation of the effects of nerve growth factor and anti-nerve growth factor on the development of central catecholaminergic neurons. Brain Res 144:277–285

Korsching S (1986) Nerve growth factor in the central nervous system. Trends Neurosci 11/12:570–573

Korsching S, Thoenen H (1983a) Nerve growth factor in sympathetic ganglia and corresponding target organs of the rat: correlation with density of sympathetic innervation. Proc Natl Acad Sci USA 80:3513–3516

Korsching S, Thoenen H (1983b) Quantitative demonstration of the retrograde axonal transport of endogenous nerve growth factor. Neurosci Lett 39:1–4

Korsching S, Thoenen H (1985a) Nerve growth factor supply for sensory neurons: site of origin and competition with the sympathetic nervous system. Neurosci Lett 54:201–205

Korsching S, Thoenen H (1985b) Treatment with 6-hydroxydopamine and colchicine decreases nerve growth factor levels in sympathetic ganglia and increases them in the corresponding target tissues. J Neurosci 5:1058–1061

Korsching S, Auburger G, Heumann R, Scott J, Thoenen H (1985) Levels of nerve growth factor and its mRNA in the central nervous system of the rat correlate with cholinergic innervation. EMBO J 4:1389–1393

Korsching S, Heumann R, Thoenen H, Hefti F (1986) Cholinergic denervation of the rat hippocampus by fimbrial transection leads to a transient accumulation of nerve growth factor (NGF) without change in mRNA^{NGF} content. Neurosci Lett 66:175–180

Levey AI, Wainer BH, Mufson EJ, Mesulam MM (1983) Co-localization of acetylcholinesterase and choline acetyltransferase in the rat cerebrum. Neuroscience 9:9–22

Levi-Montalcini R (1966) The nerve growth factor: its mode of action on sensory and sympathetic nerve cells. Harvey Lect 60:217–259

Levi-Montalcini R, Aloe L (1985) Differentiation effects of murine nerve growth factor in the peripheral and central nervous systems of Xenopus laevis tadpoles. Proc Natl Acad Sci 82:7111–7115

Levi-Montalcini R, Angeletti PU (1968) Nerve growth factor. Physiol Rev 48:534–569

Liesi P, Dahl D, Vaheri A (1983) Laminin is produced by early rat astrocytes in primary culture. J Cell Biol 96:920–924

Liesi P, Kaakkola S, Dahl D, Vaheri A (1984) Laminin is induced in astrocytes of adult brain by injury. EMBO J 3:683–686

Lindsay RM (1979) Adult rat brain astrocytes support survival of both NGF-dependent and NGF-insensitive neurones. Nature 282:80–82

Low WC, Lewis PR, Bunch ST, Dunnett SB, Thomas SR, Iversen SD, Björklund A, Stenevi U (1982) Function recovery following neural transplantation of embryonic septal nuclei in adult rats with septohippocampal lesions. Nature 300:260–262

Loy R, Moore RY (1977) Anomalous innervation of the hippocampal formation by peripheral sympathetic axons following mechanical injury. Exp Neurol 57:645–650

Lubinska L (1975) On axoplasmic flow. Int Rev Neurobiol 17:241–296

Malmfors T, Sachs C (1965) Direct studies on the disappearance of the transmitter and changes in the uptake-storage mechanisms of degenerating adrenergic nerves. Acta Physiol Scand 64:211–223

Martinez HJ, Dreyfus CF, Jonakait GM, Black IB (1985) Nerve growth factor promotes cholinergic development in brain striatal cultures. Proc Natl Acad Sci USA 82:7777–7781

Matthews DA, Nadler JV, Lynch GS, Cotman CW (1974) Development of cholinergic innervation in the hippocampal formation of the rat: I. Histochemical demonstration of acetylcholinesterase activity. Dev Biol 36:130–141

Meier R, Becker-André M, Götz R, Heumann R, Shaw A, Thoenen H (1986) Molecular cloning of bovine and chick nerve growth factor (NGF): delineation of conserved and unconserved domains and their relationship to the biological activity and antigenicity of NGF. EMBO J 5:1489–1493

Milner TA, Loy R, Amaral DG (1983) An anatomical study of the development of the septo-hippocampal projection in the rat. Dev Brain Res 8:343–371

Mobley WC, Rutkowski JL, Tennekoon GI, Buchanan K, Johnston MV (1985) Choline acetyltransferase in striatum of neonatal rats increased by nerve growth factor. Science 229:284–287

Mobley WC, Rutkowski JL, Tennekoon GI, Gemski J, Buchanan K, Johnston MV (1986) Nerve Growth Factor increases choline acetyltransferase activity in developing basal forebrain neurons. Mol Brain Res 1:53–62

Nadler JV, Matthews DA, Cotman CW, Lynch GS (1974) Development of cholinergic innervation in the hippocampal formation of the rat. Dev Biol 36:142–154

Nicoll RA (1985) The septo-hippocampal projection: a model cholinergic pathway. Trends Neurosci 8:533–536

Noda M, Ko M, Ogura A, Liu D, Amano T, Takano T, Ikawa Y (1985) Sarcoma viruses carrying *ras* oncogenes induce differentiation-associated properties in a neuronal cell line. Nature 318:73–75

O'Connor CM, Aswad DW, Clarke S (1984) Mammalian brain and erythrocyte carboxyl methyltransferases are similar enzymes that recognize both D-aspartyl and L-isoaspartyl residues in structurally altered protein substrates. Proc Natl Acad Sci USA 81:7757–7761

Ojika K, Appel S (1984) Neurotrophic effects of hippocampal extracts on medial septal nucleus *in vitro*. Proc Natl Acad Sci USA 81:2567–2571

Olson L, Ebendal T, Seiger A (1979) NGF and anti-NGF: Evidence against effects on fiber growth in locus coeruleus from cultures of perinatal central nervous system tissues. Dev Neurosci 2:160–176

Otten U (1984) Nerve growth factor and the peptidergic sensory neurons. Trends Pharmacol 7:307–310

Otten W, Weskamp G, Schlumpf M, Lichtensteiger W, Mobley WC (1985) Effects of antibodies against nerve growth factor on developing cholinergic forebrain neurons in rat. Soc Neurosci (Abstr) 11:661

Pearson RCA, Sofroniew MV, Cuello AC, Powell TPS, Eckenstein F, Esiri MM, Wilcock GK (1983) Persistence of cholinergic neurons in the basal nucleus in a brain with senile dementia of the Alzheimer's type demonstrated by immunohistochemical staining for choline acetyltransferase. Brain Res 289:375–379

Perry EK, Tomlison BE, Blessed G, Perry RH, Cross AJ, Crow TJ (1978) Correlation of cholinergic abnormalities with senile plaques and mental test scores in senile dementia. Br Med J 2:1457–1459

Price DL, Struble RG, Whitehouse PJ, Kitt CA, Cork LC (1985) Neuropathological processes in Alzheimer's disease. Drug Dev Res 5:59–67

Probstmeier R, Schachner M (1986) Epidermal growth factor is not detectable in developing and adult rodent brain by a sensitive double-site enzyme immunoassay. Neurosci Lett 63:290–294

Puma P, Buxser SE, Watson L, Kelleher DJ, Johnson GL (1983) Purification of the receptor for nerve growth factor from A875 melanoma cells by affinity chromatography. J Biol Chem 258:3370–3375

Radeke MJ, Misko TP, Hsu C, Herzenberg LA, Shooter EM (1987) Gene transfer and molecular cloning of the rat nerve growth factor receptor: a new class of receptors. Nature 325:593–597

Raivich G, Kreutzberg GW (1987) The localization and distribution of high affinity βNGF binding sites in the central nervous system of the adult rat. A light microscopic autoradiographic study using (^{125}I)βNGF. Neuroscience 20:23–36

Rennert PD, Heinrich G (1986) Nerve growth factor m RNA in brain: Localization by in situ-hybridization. Biochem Biophys Res Comm 138:813–818

Richardson PM, Verge Issa VMK, Riopelle RJ (1986) Distribution of neuronal receptors for nerve growth factor in the rat. J Neurosci 6:2312–2321

Rich KM, Yip HK, Osborne PA, Schmidt RE, Johnson EM Jr (1984) Role of nerve growth factor in the adult dorsal root ganglia neuron and its response to injury. J Comp Neurol 230:110–118

Rohrer H (1985) Nonneuronal cells from chick sympathetic and dorsal root sensory ganglion express catecholamine uptake and receptors for nerve growth factor during development. Dev Biol 111:95–107

Rossor MN, Iversen LL, Reynolds GP, Mountjoy CQ, Roth M (1984) Neurochemical characteristics of early and late onset types of Alzheimer's disease. Br Med J 288:961–968

Rush RA (1984) Immunohistochemical localization of endogenous nerve growth factor. Nature 312:364–367

Rye DB, Wainer BH, Mesulam M-M, Mufson EJ, Saper CB (1984) Cortical projections arising from the basal forebrain: A study of cholinergic and noncholinergic components employing combined retrograde tracing and immunohistochemical localization of choline acetyltransferase. Neuroscience 13:627–643

Schalling M, Hökfelt T, Goldstein M, Filer D, Yamin C, Schlesinger DH (1986) Tyrosine 3-hydroxylase in rat brain and adrenal medulla: hybridization histochemistry and immunohistochemistry combined with retrograde tracing. Proc Natl Acad Sci USA 83:6208–6212

Schwab ME, Thoenen H (1983) Retrograde axonal transport. In: Lajtha A (ed) Handbook of Neurochemistry, vol 5, Plenum Publishing Corporation, New York London, pp 381–404

Schwab M, Otten U, Agid Y, Thoenen H (1979) Nerve growth factor (NGF) in the rat CNS: absence of specific retrograde axonal transport and tyrosine hydroxylase induction in locus coeruleus and substantia nigra. Brain Res 168:473–483

Schwab ME, Heumann R, Thoenen H (1982) Communication between target organs and nerve cells: retrograde axonal transport and site of action of nerve growth factor. Cold Spring Habor Symposia on Quantitative Biology, vol 46, Cold Spring Harbor Laboratory, pp 125–134

Scott J, Selby M, Urdea M, Quiroga M, Bell GI, Rutter WJ (1983) Isolation and nucleotide sequence of a cDNA encoding the precursor of mouse nerve growth factor. Nature 302:538–540

Seeley PJ, Rukenstein A, Connolly JL, Greene LA (1984) Differential inhibition of nerve growth factor and epidermal growth factor effects on the PC12 pheochromocytoma line. J Cell Biol 98:417–426

Seiler M, Schwab ME (1984) Specific retrograde transport of nerve growth factor (NGF) from neocortex to nucleus basalis in the rat. Brain Res 300:33–39

Shelton DL, Reichardt LF (1984) Expression of the nerve growth factor gene correlates with the density of sympathetic innervation in effector organs. Proc Natl Acad Sci USA 81:7951–7955

Shelton DL, Reichardt LF (1986a) Studies on the expression of the β nerve growth factor (NGF) gene in the central nervous system: Level and regional distribution of NGF mRNA suggest that NGF functions as a trophic factor for several distinct populations of neurons. Proc Natl Acad Sci USA 83:2714–2718

Shelton DL, Reichardt LF (1986b) Studies on the regulation of beta-nerve growth factor gene expression in the rat iris: The level of mRNA-encoding nerve growth factor is increased in irises placed in explant cultures in vitro, but not in irises deprived of sensory or sympathetic innervation in vivo. J Cell Biol 102:1940–1948

Sofroniew MV, Eckenstein F, Thoenen H, Cuello AC (1982) Topography of choline acetyltransferase-containing neurons in the forebrain of the rat. Neurosci Lett 33:7–12

Sorimachi M, Kataoka K (1975) High affinity choline uptake: an early index of cholinergic innervation in rat brain. Brain Res 94:325–336

Springer JE, Loy R (1985) Intrahippocampal injections of antiserum to nerve growth factor inhibit sympathohippocampal sprouting. Brain Res Bull 15:629–634

Stein DG, Will BE (1983) Nerve growth factor produces a temporary facilitation of recovery from entorhinal cortex lesions. Brain Res 261:127–131

Stenevi U, Björklund A (1978) Growth of vascular sympathetic axons into the hippocampus after lesions of the septo-hippocampal pathway: a pitfall in brain lesion studies. Neurosci Lett 7:219–224

Sutter A, Hosang M, Vale RD, Shooter EM (1984) The interaction of nerve growth factor with its specifric receptors. In: Black IB (ed) Cellular and molecular biology of neuronal development. Plenum Press, New York, pp 201–214

Taniuchi M, Clark HB, Johnson EM Jr (1986) Induction of nerve growth factor receptor in Schwann cells after axotomy. Proc Natl Acad Sci USA 83:4094–4098

Thoenen H, Barde YA (1980) Physiology of nerve growth factor. Physiol Rev 60:1284–1335

Thoenen H, Edgar D (1985) Neurotrophic factors. Science 229:238–242

Thoenen H, Korsching S, Heumann R, Acheson A (1985) Nerve Growth Factor. In: Growth factors in biology and medicine. Pitman, London (Ciba Foundation Symposium 116) pp 113–128

Thoenen H, Auburger G, Hellweg R, Heumann R, Korsching S (1987a) Cholinergic innervation and levels of nerve growth factor and its mRNA in the central nervous system. In: Cholinergic Mechanisms, vol. 6, Cellular and Molecular Basis of Cholinergic Function, Ellis Horwood, Chichester (in press)

Thoenen H, Barde Y-A, Davies AM, Johnson JE (1987b) Neurotrophic factors and neuronal death. In: Selective neuronal death. Wiley, Chichester (Ciba Foundation Symposium 126) pp 82–95

Ullrich A, Gray A, Berman C, Dull TJ (1983) Human β-nerve growth factor gene sequence highly homologous to that of mouse. Nature 303:821–825

Wakade AR, Edgar D, Thoenen H (1983) Both nerve growth factor and high K^+ concentrations support the survival of chick embryo sympathetic neurons. Exp Cell Res 144:377–384

Whittemore SR, Ebendal T, Lärkfors L, Olson L, Seiger Å, Strömberg I, Persson H (1986) Developmental and regional expression of β nerve growth factor messenger RNA and protein in the rat central nervous system. Proc Natl Acad Sci USA 83:817–821

Wilcock GK, Esiri MM, Bowen DM, Smith CCT (1983) The nucleus basalis in Alzheimer's disease: cell counts and cortical biochemistry. Neuropathol Appl Neurobiol 9:175–179

Will B, Hefti F (1985) Behavioural and neurochemical effects of chronic intraventricular injections of nerve growth factor in adult rats with fimbria lesions. Behav Brain Res 17:17–24

Williams LR, Varon S, Peterson GM, Wictorin K, Fischer W, Björklund A, Gage FH (1986) Continuous infusion of nerve growth factor prevents basal forebrain neuronal death after fimbria fornix transection. Proc Natl Acad Sci USA 83:9231–9235

Wion D, Dicou E, Brachet P (1984) Synthesis and partial maturation of the α- and γ-subunits of the mouse submaxillary gland nerve growth factor in *Xenopus laevis* oocytes. FEBS Lett 166:104–108

Wisniewski HM, Merz GS (1983) Neuritic and amyloid plaques in senile dementia of the Alzheimer type. In: Katzman R (ed) Banbury Report 15: Biological aspects of Alzheimer's diseases. Cold Spring Harbor Laboratory, Cold Spring Harbor, pp 145–153

Zimmer J, Haug F-M S (1978) Laminar differentiation of the hippocampus, fascia dentata and subiculum in developing rats, observed with the Timm sulphide silver method. J Comp Neurol 179:581–618

Zimmermann A, Sutter A (1983) β-nerve factor (β-NGF) receptors on glial-cells – cell-cell interaction between neurons and Schwann-cells in cultures of chick sensory ganglia. EMBO J 2:879–885

Subject Index